AF885676

CONTENTS

The College Board is a national nonprofit membership association dedicated to preparing, inspiring, and connecting students to college and opportunity. Founded in 1900, the association is composed of nearly 4,000 schools, colleges, universities, and other educational organizations. Each year, the College Board serves over three million students and their parents, 22,000 high schools, and 3,500 colleges, through major programs and services in college admission, guidance, assessment, financial aid, enrollment, and teaching and learning. Among its best-known programs are the SAT®, the PSAT/NMSQT™, the Advanced Placement Program® (AP®), and Pacesetter®. The College Board is committed to the principles of equity and excellence, and that commitment is embodied in all of its programs, services, activities, and concerns.

PREFACE

In order to learn biology and enjoy it as a science, you need to engage in the processes of science. Recent surveys have shown that most colleges and universities have a lab component in their first-year general biology course. The AP Biology course is equivalent to a college-level course both in quality and sophistication, and therefore has a substantial lab component.

The AP Biology Development Committee has produced this set of 12 labs to provide a framework for your lab experience. In addition to the specific objectives, the labs are designed to help you learn how to:

- understand problems;
- develop hypotheses;
- design and implement controlled experiments;
- identify independent and dependent variables;
- analyze data;
- draw conclusions;
- think analytically; and
- communicate results with appropriate data tables and graphs.

In these labs, you are challenged to do experiments drawn from some of the more important areas within modern biology. These include biological chemistry, cell structure and function, energy transformation, molecular genetics, heredity, evolution, plant structure and physiology, animal structure and physiology, behavior, and ecology.

We hope these lab exercises will serve as basic introductions, or springboards, into further experiments, studies, or independent projects.

The AP Biology Development Committee

NOTE TO STUDENTS: You should perform the experiments and procedures described in this manual following the directions given and only under the supervision of a qualified instructor. It is the instructor's responsibility to determine what safety precautions you should take in addition to those indicated in the manual. Should an accident occur during any of these procedures, stop work immediately and notify the instructor.

LAB ONE DIFFUSION AND OSMOSIS

OVERVIEW

In this lab you will:

1. investigate the processes of diffusion and osmosis in a model membrane system, and
2. investigate the effect of solute concentration on water potential as it relates to living plant tissues.

OBJECTIVES

Before doing this lab you should understand:

- the mechanisms of diffusion and osmosis and their importance to cells;
- the effects of solute size and concentration gradients on diffusion across selectively permeable membranes;
- the effects of a selectively permeable membrane on diffusion and osmosis between two solutions separated by the membrane;
- the concept of water potential;
- the relationship between solute concentration and pressure potential and the water potential of a solution; and
- the concept of molarity and its relationship to osmotic concentration.

After doing this lab you should be able to:

- measure the water potential of a solution in a controlled experiment;
- determine the osmotic concentration of living tissue *or* an unknown solution from experimental data;
- describe the effects of water gain or loss in animal and plant cells; and
- relate osmotic potential to solute concentration and water potential.

INTRODUCTION

Many aspects of the life of a cell depend on the fact that atoms and molecules have kinetic energy and are constantly in motion. This kinetic energy causes molecules to bump into each other and move in new directions. One result of this molecular motion is the process of diffusion.

Diffusion is the random movement of molecules from an area of higher concentration of those molecules to an area of lower concentration. For example, if one were to open a bottle of hydrogen sulfide (H_2S has the odor of rotten eggs) in one corner of a room, it would not be long before someone in the opposite corner would perceive the smell of rotten eggs. The bottle contains a higher concentration of H_2S molecules than the room does and therefore the H_2S gas diffuses from the area of higher concentration to the area of lower concentration. Eventually, a dynamic equilibrium will be reached; the concentration of H_2S will be approximately equal throughout the room and no **net** movement of H_2S will occur from one area to the other.

Osmosis is a special case of diffusion. Osmosis is the diffusion of water through a selectively permeable membrane (a membrane that allows for diffusion of certain solutes and water) from a region of higher water potential to a region of lower water potential. Water potential is the measure of free energy of water in a solution.

Diffusion and osmosis do not entirely explain the movement of ions or molecules into and out of cells. One property of a living system is active transport. This process uses energy from ATP to move substances through the cell membrane. Active transport usually moves substances against a concentration gradient, from regions of low concentration of that substance into regions of higher concentration.

EXERCISE 1A: Diffusion

In this experiment you will measure diffusion of small molecules through dialysis tubing, an example of a selectively permeable membrane. Small solute molecules and water molecules can move freely through a selectively permeable membrane, but larger molecules will pass through more slowly, or perhaps not at all. The movement of a solute through a selectively permeable membrane is called **dialysis.** The size of the minute pores in the dialysis tubing determines which substances can pass through the membrane.

A solution of glucose and starch will be placed inside a bag of dialysis tubing. Distilled water will be placed in a beaker, outside the dialysis bag. After 30 minutes have passed, the solution inside the dialysis tubing and the solution in the beaker will be tested for glucose and starch. The presence of glucose will be tested with Benedict's solution, Testape®, or Clinistix®. The presence of starch will be tested with Lugol's solution (Iodine Potassium-Iodide, or IKI).

Procedure

1. Obtain a 30-cm piece of 2.5-cm dialysis tubing that has been soaking in water. Tie off one end of the tubing to form a bag. To open the other end of the bag, rub the end between your fingers until the edges separate.
2. Test the 15% glucose/1% starch solution for the presence of glucose. Your teacher may have you do a Benedict's test or use glucose Testape® or Clinistix®. Record the results in Table 1.1.
3. Place 15 mL of the 15% glucose/1% starch solution in the bag. Tie off the other end of the bag, leaving sufficient space for the expansion of the contents in the bag. Record the color of the solution in Table 1.1.
4. Fill a 250-mL beaker or cup two-thirds full with distilled water. Add approximately 4 mL of Lugol's solution to the distilled water and record the color of the solution in Table 1.1. Test this solution for glucose and record the results in Table 1.1.
5. Immerse the bag in the beaker of solution.
6. Allow your setup to stand for approximately 30 minutes or until you see a distinct color change in the bag or in the beaker. Record the final color of the solution in the bag, and of the solution in the beaker, in Table 1.1.
7. Test the liquid in the beaker and in the bag for the presence of glucose. Record the results in Table 1.1.

Table 1.1

	Initial Contents	Solution Color		Presence of Glucose	
		Initial	Final	Initial	Final
Bag	15% glucose & 1% starch				
Beaker	H_2O & IKI				

Analysis of Results

1. Which substance(s) are entering the bag and which are leaving the bag? What experimental evidence supports your answer?

2. Explain the results you obtained. Include the concentration differences and membrane pore size in your discussion.

3. Quantitative data uses numbers to measure observed changes. How could this experiment be modified so that quantitative data could be collected to show that water diffused into the dialysis bag?

4. Based on your observations, rank the following by relative size, beginning with the smallest: glucose molecules, water molecules, IKI molecules, membrane pores, starch molecules.

5. What results would you expect if the experiment started with a glucose and IKI solution inside the bag and only starch and water outside? Why?

____

____

____

____

____

____

EXERCISE 1B: Osmosis

In this experiment you will use dialysis tubing to investigate the relationship between solute concentration and the movement of water through a selectively permeable membrane by the process of osmosis.

When two solutions have the same concentration of solutes, they are said to be **isotonic** to each other (*iso-* means same, *-ton* means condition, *-ic* means pertaining to). If the two solutions are separated by a selectively permeable membrane, water will move between the two solutions, but there will be no **net** change in the amount of water in either solution.

If two solutions differ in the concentration of solutes that each has, the one with more solute is **hypertonic** to the one with less solute (*hyper-* means over, or more than). The solution that has less solute is **hypotonic** to the one with more solute (*hypo-* means under, or less than). These words can only be used to compare solutions.

Now consider two solutions separated by a selectively permeable membrane. The solution that is hypertonic to the other must have more solute and therefore less water. At standard atmospheric pressure, the water potential of the hypertonic solution is less than the water potential of the hypotonic solution, so the **net** movement of water will be from the hypotonic solution into the hypertonic solution.

Label the sketch in Figure 1.1 to indicate which solution is hypertonic and which is hypotonic, and use arrows to show the initial **net** movement of water.

Figure 1.1

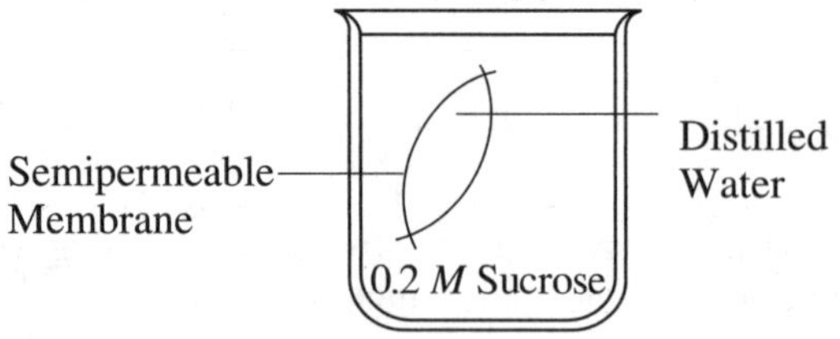

Procedure

1. Obtain six 30-cm strips of presoaked dialysis tubing.
2. Tie a knot in one end of each piece of dialysis tubing to form 6 bags. Pour approximately 15–25 mL of each of the following solutions into separate bags:

   a) distilled water
   b) 0.2 *M* sucrose
   c) 0.4 *M* sucrose
   d) 0.6 *M* sucrose
   e) 0.8 *M* sucrose
   f) 1.0 *M* sucrose

   Remove most of the air from each bag by drawing the dialysis bag between two fingers. Tie off the other end of the bag. Leave sufficient space for the expansion of the contents in the bag. (The solution should fill only about one-third to one-half of the piece of tubing.)
3. Rinse each bag gently with distilled water to remove any sucrose spilled during the filling.
4. Carefully blot the outside of each bag and record in Table 1.2 the initial mass of each bag, expressed in grams.
5. Place each bag in an empty 250-mL beaker or cup and label the beaker to indicate the molarity of the solution in the dialysis bag.
6. Now fill each beaker two-thirds full with distilled water. *Be sure to completely submerge each bag.*
7. Let them stand for 30 minutes.
8. At the end of 30 minutes remove the bags from the water. Carefully blot and determine the mass of each bag.
9. Record your group's data in Table 1.2. Obtain data from the other lab groups in your class to complete Table 1.3 on page 6.

Table 1.2: Dialysis Bag Results—Group Data

Contents in Dialysis Bag	Initial Mass	Final Mass	Mass Difference	Percent Change in Mass*
a) 0.0 *M* Distilled Water				
b) 0.2 *M* Sucrose				
c) 0.4 *M* Sucrose				
d) 0.6 *M* Sucrose				
e) 0.8 *M* Sucrose				
f) 1.0 *M* Sucrose				

* To calculate:

$$\textbf{Percent Change in Mass} = \frac{\textbf{Final Mass} - \textbf{Initial Mass}}{\textbf{Initial Mass}} \times \textbf{100}$$

Table 1.3: Dialysis Bag Results—Class Data

	Percent Change in Mass of Dialysis Bags									
	Group 1	Group 2	Group 3	Group 4	Group 5	Group 6	Group 7	Group 8	Total	Class Average
0.0 *M* Distilled Water										
0.2 *M* Sucrose										
0.4 *M* Sucrose										
0.6 *M* Sucrose										
0.8 *M* Sucrose										
1.0 *M* Sucrose										

10. Graph the results for both your individual data and the class average on Graph 1.1. *

For this graph you will need to determine the following:

a. The *independent* variable: ______________________________
Use this to label the horizontal (x) axis.

b. The *dependent* variable: ______________________________
Use this to label the vertical (y) axis

Graph 1.1 **Title:** ____

* See Appendix II for information on graphing.

Analysis of Results

1. Explain the relationship between the change in mass and the molarity of sucrose within the dialysis bags.

2. Predict what would happen to the mass of each bag in this experiment if all the bags were placed in a 0.4 *M* sucrose solution instead of distilled water. Explain your response.

3. Why did you calculate the percent change in mass rather than simply using the change in mass?

4. A dialysis bag is filled with distilled water and then placed in a sucrose solution. The bag's initial mass is 20 g and its final mass is 18 g. Calculate the percent change of mass, showing your calculations here.

5. The sucrose solution in the beaker would have been ____________ to the distilled water in the bag. (Circle the word that best completes the sentence.)

isotonic **hypertonic** **hypotonic**

EXERCISE 1C: Water Potential

In this part of the exercise you will use potato cores placed in different molar concentrations of sucrose in order to determine the water potential of potato cells. First, however, we will explore what is meant by the term "water potential."

Botanists use the term water potential when predicting the movement of water into or out of plant cells. Water potential is abbreviated by the Greek letter psi (ψ) and it has two components: a physical pressure component (pressure potential ψ_p) and the effects of solutes (solute potential ψ_s).

$$\psi = \psi_p + \psi_s$$

Water potential = **Pressure potential** + **Solute potential**

Water will always move from an area of higher water potential (higher free energy; more water molecules) **to an area of lower water potential** (lower free energy; fewer water molecules). Water potential, then, measures the tendency of water to leave one place in favor of another place. You can picture the water diffusing "down" a water potential gradient.

Water potential is affected by two physical factors. One factor is the addition of solute which lowers the water potential. The other factor is pressure potential (physical pressure). An increase in pressure raises the water potential. By convention, the water potential of pure water at atmospheric pressure is defined as being zero ($\psi = 0$). For instance, it can be calculated that a 0.1-*M* solution of sucrose at atmospheric pressure ($\psi_p = 0$) has a water potential of -2.3 bars due to the solute ($\psi = -2.3$).*

Movement of H_2O into and out of a cell is influenced by the solute potential (relative concentration of solute) on either side of the cell membrane. If water moves out of the cell, the cell will shrink. If water moves into an animal cell, it will swell and may even burst. In plant cells, the presence of a cell wall prevents cells from bursting as water enters the cells, but pressure eventually builds up inside the cell and affects the net movement of water. As water enters a dialysis bag or a cell with a cell wall, pressure will develop inside the bag or cell as water pushes against the bag or cell wall. The pressure would cause, for example, the water to rise in an osmometer tube or increase the pressure on a cell wall. It is important to realize that water potential and solute concentration are inversely related. The addition of solutes lowers the water potential of the system. In summary, solute potential is the effect that solutes have on a solution's overall water potential.

Movement of H_2O into and out of a cell is also influenced by the pressure potential (physical pressure) on either side of the cell membrane. Water movement is directly proportional to the pressure on a system. For example, pressing on the plunger of a water-filled syringe causes the water to exit via any opening. In plant cells this physical pressure can be exerted by the cell pressing against the partially elastic cell wall. Pressure potential is usually positive in living cells; in dead xylem elements it is often negative.

---

* A bar is a metric measure of pressure, measured with a barometer, that is about the same as 1 atmosphere. Another measure of pressure is the megapascal (MPa). [1 MPa = 10 bars.]

It is important for you to be clear about the numerical relationships between water potential and its components, pressure potential and solute potential. The water potential value can be positive, zero, or negative. Remember that water will move across a membrane in the direction of the lower water potential. An increase in pressure potential results in a more positive value, and a decrease in pressure potential (tension or pulling) results in a more negative value. In contrast to pressure potential, solute potential is always negative; since pure water has a water potential of zero, any solutes will make the solution have a lower (more negative) water potential. Generally, an increase in solute potential makes the water potential value more negative and an increase in pressure potential makes the water potential more positive.

To illustrate the concepts discussed above, we will look at a sample system using Figure 1.2. When a solution, such as that inside a potato cell, is separated from pure water by a selectively permeable cell membrane, water will move (by osmosis) from the surrounding water where water potential is higher, into the cell where water potential is lower (more negative) due to the solute potential (ψ_s). In Figure 1.2a the pure water potential (ψ) is 0 and the solute potential (ψ_s) is -3. We will assume, for purposes of explanation, that the solute is not diffusing out of the cell. By the end of the observation, the movement of water into the cell causes the cell to swell and the cell contents to push against the cell wall to produce an increase in pressure potential (turgor) ($\psi_p = 3$). Eventually, enough turgor pressure builds up to balance the negative solute potential of the cell. When the water potential of the cell equals the water potential of the pure water outside the cell (ψ of cell = ψ of pure water = 0), a dynamic equilibrium is reached and there will be no *net* water movement (Figure 1.2b).

Figure 1.2

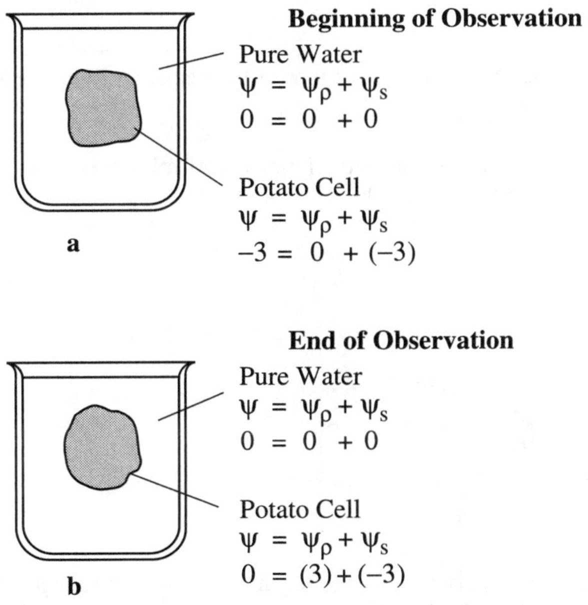

If you were to add solute to the water outside the potato cells, the water potential of the solution surrounding the cells would decrease. It is possible to add just enough solute to the water so that the water potential outside the cell is the same as the water potential inside the cell. In this case, there will be no net movement of water. This does not mean, however, that the solute concentrations inside and outside the cell are equal, because water potential inside the cell results from the combination of both pressure potential and solute potential (Figure 1.3).

Figure 1.3

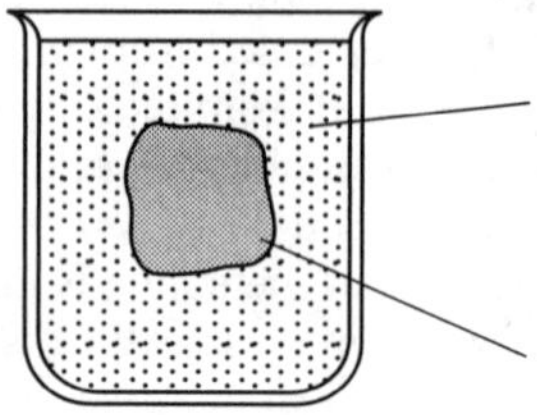

If enough solute is added to the water outside the cells, water will leave the cells, moving from an area of higher water potential to an area of lower water potential. The loss of water from the cells will cause the cells to lose turgor. A continued loss of water will eventually cause the cell membrane to shrink away from the cell wall (plasmolysis).

Procedure

Work in groups. You will be assigned one or more of the beaker contents listed in Table 1.4. For each of these, do the following:

1. Pour 100 mL of the assigned solution into a labeled 250-mL beaker. Slice a potato into discs that are approximately 3 cm thick (see Figure 1.4).

Figure 1.4

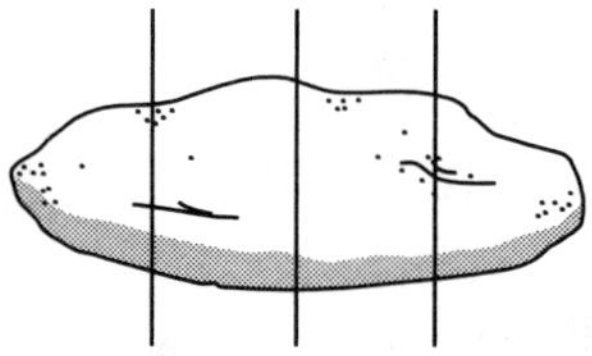

2. Use a cork borer (approximately 5 mm in inner diameter) to cut four potato cylinders. Do not include any skin on the cylinders. You need four potato cylinders for *each* beaker.
3. Keep your potato cylinders in a covered beaker until it is your turn to use the balance.
4. Determine the mass of the four cylinders together and record the mass in Table 1.4. Put the four cylinders into the beaker of sucrose solution.
5. Cover the beaker with plastic wrap to prevent evaporation.
6. Let it stand overnight.
7. Remove the cores from the beakers, blot them gently on a paper towel, and determine their total mass.
8. Record the final mass in Table 1.4 and record class data in Table 1.5. Calculate the percentage change as you did in Exercise 1B. Do this for both your individual results and the class average.
9. Graph both your individual data and the class average for the percentage change in mass in Table 1.4.

Table 1.4: Potato Core—Individual Data

Contents in Beaker	Initial Mass	Final Mass	Mass Difference	Percent Change in Mass	Class Average Percent Change in Mass
a) 0.0 *M* Distilled Water					
b) 0.2 *M* Sucrose					
c) 0.4 *M* Sucrose					
d) 0.6 *M* Sucrose					
e) 0.8 *M* Sucrose					
f) 1.0 *M* Sucrose					

Table 1.5: Potato Core Results—Class Data

Contents in Beaker	Percent Change in Mass of Potato Cores								Total	Class Average
	Group 1	Group 2	Group 3	Group 4	Group 5	Group 6	Group 7	Group 8		
0.0 *M* Distilled Water										
0.2 *M* Sucrose										
0.4 *M* Sucrose										
0.6 *M* Sucrose										
0.8 *M* Sucrose										
1.0 *M* Sucrose										

Graph 1.2: Percent Change in Mass of Potato Cores at Different Molarities of Sucrose

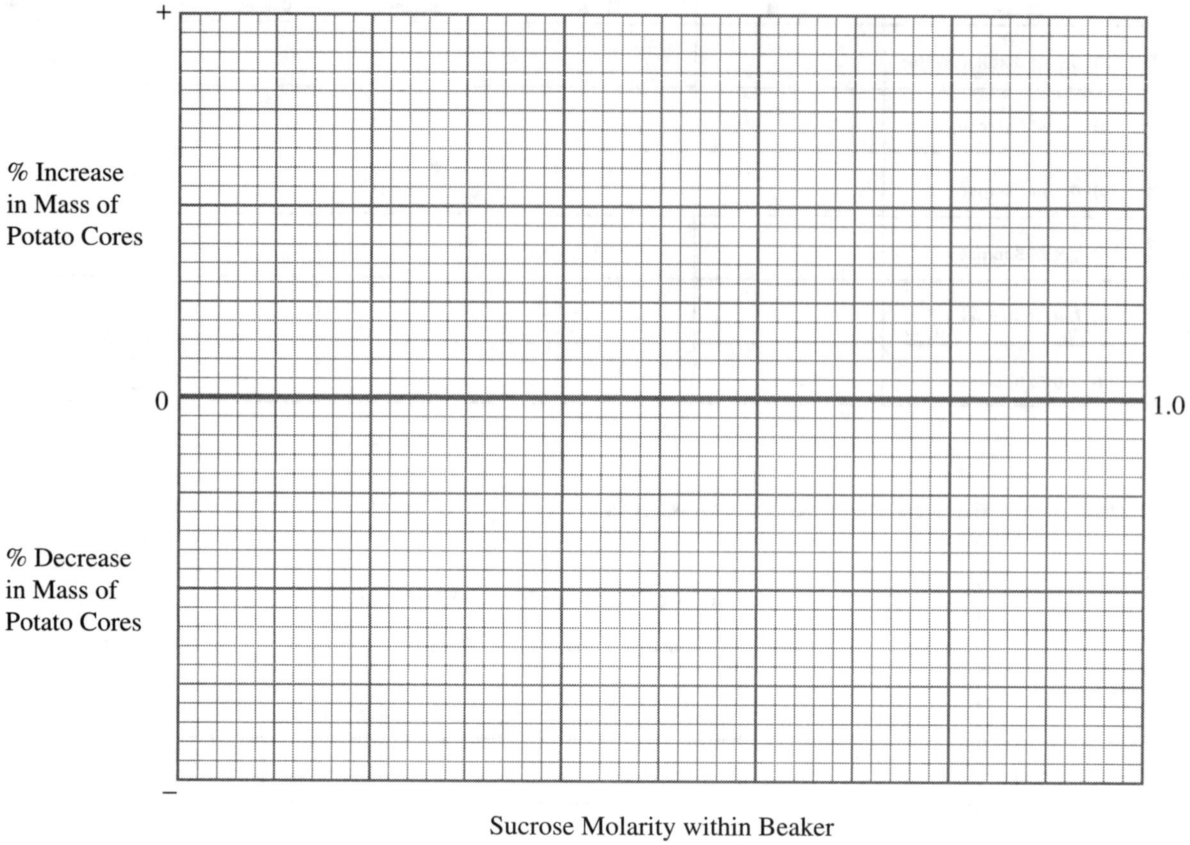

10. Determine the molar concentration of the potato core. This would be the sucrose molarity in which the mass of the potato core does not change. To find this, follow your teacher's directions to draw the straight line on Graph 1.2 that best fits your data. **The point at which this line crosses the x-axis represents the molar concentration of sucrose with a water potential that is equal to the potato tissue water potential.** At this concentration there is no net gain or loss of water from the tissue. Indicate this concentration of sucrose in the space provided below.

Molar concentration of sucrose = _______________ *M*

EXERCISE 1D: Calculation of Water Potential from Experimental Data

1. The solute potential of this sucrose solution can be calculated using the following formula:

$$\psi_s = -\mathbf{iCRT}$$

where **i = Ionization constant** (for sucrose this is 1.0 because sucrose does not ionize in water)

C = Molar concentration (determined above)

R = Pressure constant (R = 0.0831 liter bars/mole °K)

T = Temperature °K (273 + °C of solution)

The units of measure will cancel as in the following example:

A 1.0 *M* sugar solution at 22°C under standard atmospheric conditions

$$\psi_s = -\mathbf{i} \times \mathbf{C} \times \mathbf{R} \times \mathbf{T}$$

ψ_s = −(1)(1.0 mole/liter)(0.0831 liter bar/mole °K)(295 °K)
ψ_s = −24.51 bars

2. Knowing the solute potential of the solution (ψ_s) and knowing that the pressure potential of the solution is zero ($\psi_p = 0$) allows you to calculate the water potential of the solution. The water potential will be equal to the solute potential of the solution.

$$\psi = 0 + \psi_s \text{ or } \psi = \psi_s$$

The water potential of the solution at equilibrium will be equal to the water potential of the potato cells. What is the water potential of the potato cells? Show your calculations here:

3. Water potential values are useful because they allow us to predict the direction of the flow of water. Recall from the discussion that water flows from an area of higher water potential to an area of lower water potential.

For the sake of discussion, suppose that a student calculates that the water potential of a solution inside a bag is −6.25 bar ($\psi_s = -6.25$, $\psi_p = 0$) and the water potential of a solution surrounding the bag is −3.25 bar ($\psi_s = -3.25$, $\psi_p = 0$). In which direction will the water flow?

Water will flow into the bag. This occurs because there are more solute molecules inside the bag (therefore a value further away from zero) than outside in the solution.

Questions

1. If a potato core is allowed to dehydrate by sitting in the open air, would the water potential of the potato cells decrease or increase? Why?

____

____

____

____

____

____

2. If a plant cell has a lower water potential than its surrounding environment and if pressure is equal to zero, is the cell hypertonic (in terms of solute concentration) or hypotonic to its environment? Will the cell gain water or lose water? Explain your response.

____

____

____

____

____

____

Figure 1.5

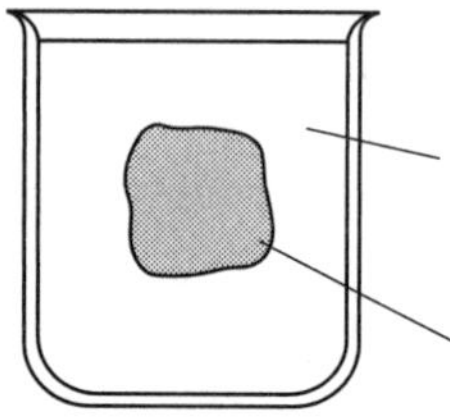

Initial Values Beaker Contents

(0.4% sucrose solution) $\psi_s = (-)4$

Dialysis Bag

(with 0.1% sucrose solution)

$\psi_s = (-)1,\ \psi_p = 0$

3. In Figure 1.5 the beaker is open to the atmosphere. What is the pressure potential (ψ_p) of the system?

____

4. In Figure 1.5 where is the greatest water potential? (Circle one.)

beaker **dialysis bag**

5. Water will diffuse ____________ (circle one) the bag. Why?

into **out of**

____

____

6. Zucchini cores placed in sucrose solutions at 27°C resulted in the following percent changes after 24 hours:

% Change in Mass	Sucrose Molarity
20%	Distilled Water
10%	0.2 *M*
– 3%	0.4 *M*
–17%	0.6 *M*
–25%	0.8 *M*
–30%	1.0 *M*

7. **a.** Graph the results on Graph 1.3.

Graph 1.3 **Title:** ____________________

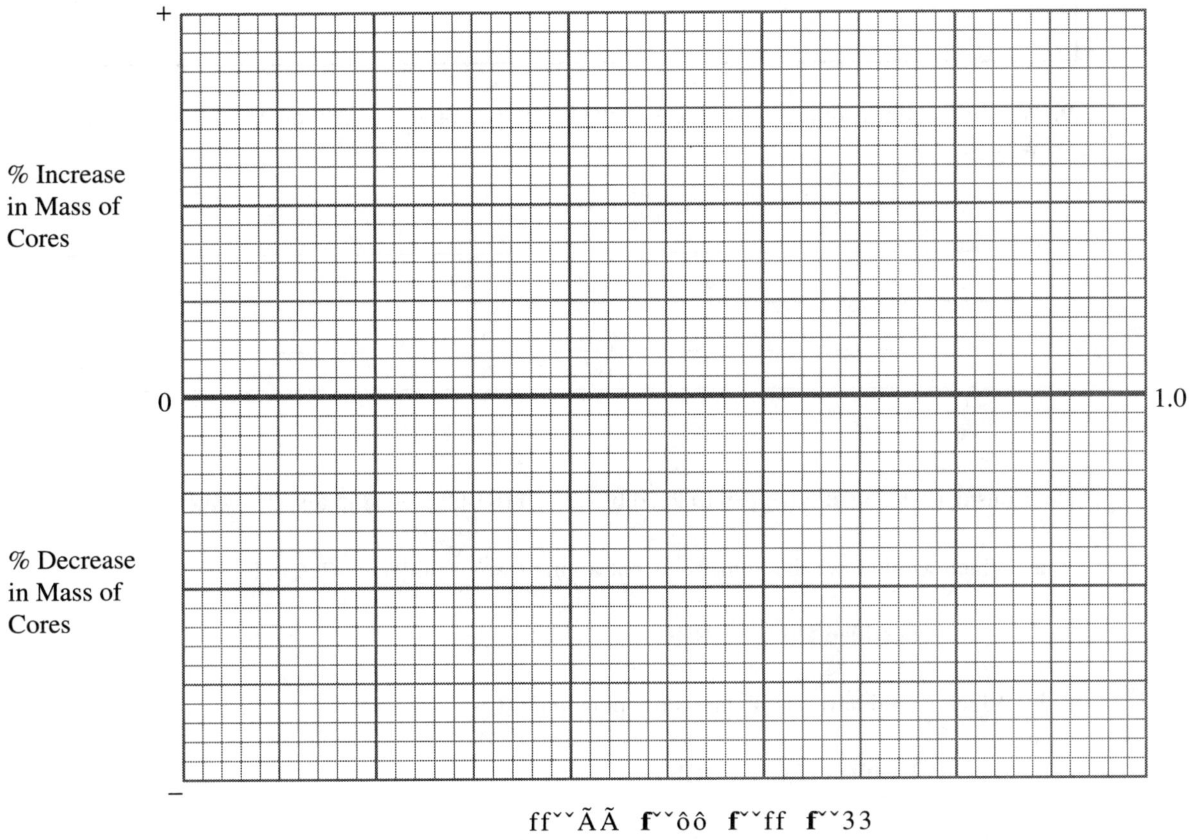

b. What is the molar concentration of solutes within the zucchini cells? ____________________

8. Refer to the procedure for calculating water potential from experimental data (page 13).

a. Calculate solute potential (ψ_s) of the sucrose solution in which the mass of the zucchini cores does not change. Show your work here:

b. Calculate the water potential (ψ) of the solutes within the zucchini cores. Show your work here:

9. What effect does adding solute have on the solute potential component (ψ_s) of that solution? Why?

10. Consider what would happen to a red blood cell (RBC) placed in distilled water:

a. Which would have the higher concentration of water molecules? (Circle one.)

Distilled H_2O **RBC**

b. Which would have the higher water potential? (Circle one.)

Distilled H_2O **RBC**

c. What would happen to the red blood cell? Why?

EXERCISE 1E: Onion Cell Plasmolysis

Plasmolysis is the shrinking of the cytoplasm of a plant cell in response to diffusion of water out of the cell and into a hypertonic solution (high solute concentration) surrounding the cell as shown in Figure 1.6. During plasmolysis the cellular membrane pulls away from the cell wall. In the next lab exercise you will examine the details of the effects of highly concentrated solutions on diffusion and cellular contents.

Figure 1.6

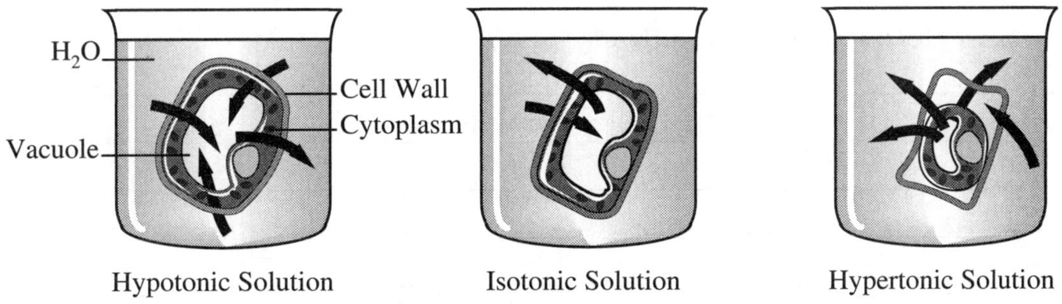

Procedure

1. Prepare a wet mount of a small piece of the epidermis of an onion. Observe under 100X magnification. Sketch and describe the appearance of the onion cells.

____

____

____

2. Add 2 or 3 drops of 15% NaCl to one edge of the cover slip. Draw this salt solution across the slide by touching a piece of paper towel to the fluid under the opposite edge of the cover slip. Sketch and describe the onion cells. Explain what has happened.

____

____

____

3. Remove the cover slip and flood the onion epidermis with fresh water. Observe under 100X. Describe and explain what happened.

____

____

____

Analysis of Results

1. What is plasmolysis?

2. Why did the onion cells plasmolyze?

3. In the winter, grass often dies near roads that have been salted to remove ice. What causes this to happen?

LAB TWO ENZYME CATALYSIS

OVERVIEW

In this lab you will:

1. observe the conversion of hydrogen peroxide (H_2O_2) to water and oxygen gas by the enzyme catalase, and
2. measure the amount of oxygen generated and calculate the rate of the enzyme-catalyzed reaction.

OBJECTIVES

Before doing this lab you should understand:

- the general functions and activities of enzymes;
- the relationship between the structure and function of enzymes;
- the concept of initial reaction rates of enzymes;
- how the concept of free energy relates to enzyme activity;
- that changes in temperature, pH, enzyme concentration, and substrate concentration can affect the initial reaction rates of enzyme-catalyzed reactions; and
- catalyst, catalysis, and catalase.

After doing this lab you should be able to:

- measure the effects of changes in temperature, pH, enzyme concentration, and substrate concentration on reaction rates of an enzyme-catalyzed reaction in a controlled experiment; and
- explain how environmental factors affect the rate of enzyme-catalyzed reactions.

INTRODUCTION

In general, **enzymes** are proteins produced by living cells; they act as catalysts in biochemical reactions. A **catalyst** affects the rate of a chemical reaction. One consequence of enzyme activity is that cells can carry out complex chemical activities at relatively low temperatures.

In an enzyme-catalyzed reaction, the substance to be acted upon, the **substrate (S),** binds reversibly to the active site of the **enzyme (E).** One result of this temporary union is a reduction in the energy required to activate the reaction of the substrate molecule so that the **products (P)** of the reaction are formed. In summary:

$$\mathbf{E + S \rightarrow ES \rightarrow E + P}$$

Note that the enzyme is not changed in the reaction and can be recycled to break down additional substrate molecules. Each enzyme is specific for a particular reaction because its amino acid sequence is unique and causes it to have a unique three-dimensional structure. The **active site** is the portion of the enzyme that interacts with the substrate, so that any substance that blocks or changes the shape of the active site affects the activity of the enzyme. A description of several ways enzyme action may be affected follows:

1. **Salt Concentration.** If the salt concentration is close to zero, the charged amino acid side chains of the enzyme molecules will attract each other. The enzyme will denature and form an inactive precipitate. If, on the other hand, the salt concentration is very high, normal interaction of charged groups will be blocked, new interactions will occur, and again the enzyme will precipitate. An intermediate salt concentration, such as that of human blood (0.9%) or cytoplasm, is the optimum for many enzymes.

2. **pH.** pH is a logarithmic scale that measures the acidity, or H^+ concentration, in a solution. The scale runs from 0 to 14 with 0 being highest in acidity and 14 lowest. When the pH is in the range of 0–7, a solution is said to be acidic; if the pH is around 7, the solution is neutral; and if the pH is in the range of 7–14, the solution is basic. Amino acid side chains contain groups, such as $-COOH$ and $-NH_2$, that readily gain or lose H^+ ions. As the pH is lowered an enzyme will tend to gain H^+ ions, and eventually enough side chains will be affected so that the enzyme's shape is disrupted. Likewise, as the pH is raised, the enzyme will lose H^+ ions and eventually lose its active shape. Many enzymes perform optimumly in the neutral pH range and are denatured at either an extremely high or low pH. Some enzymes, such as pepsin, which acts in the human stomach where the pH is very low, have a low pH optimum.

3. **Temperature.** Generally, chemical reactions speed up as the temperature is raised. As the temperature increases, more of the reacting molecules have enough kinetic energy to undergo the reaction. Since enzymes are catalysts for chemical reactions, enzyme reactions also tend to go faster with increasing temperature. However, if the temperature of an enzyme-catalyzed reaction is raised still further, a **temperature optimum** is reached; above this value the kinetic energy of the enzyme and water molecules is so great that the conformation of the enzyme molecules is disrupted. The positive effect of speeding up the reaction is now more than offset by the negative effect of changing the conformation of more and more enzyme molecules. Many proteins are denatured by temperatures around 40–50°C, but some are still active at 70–80°C, and a few even withstand boiling.

4. **Activations and Inhibitors.** Many molecules other than the substrate may interact with an enzyme. If such a molecule increases the rate of the reaction it is an **activator,** and if it decreases the reaction rate it is an **inhibitor.** These molecules can regulate how fast the enzyme acts. Any substance that tends to unfold the enzyme, such as an organic solvent or detergent, will act as an inhibitor. Some inhibitors act by reducing the -S-S- bridges that stabilize the enzyme's structure. Many inhibitors act by reacting with side chains in or near the active site to change its shape or block it. Many well-known poisons, such as potassium cyanide and curare, are enzyme inhibitors that interfere with the active site of critical enzymes.

The enzyme used in this lab, catalase, has four polypeptide chains, each composed of more than 500 amino acids. This enzyme is ubiquitous in aerobic organisms. One function of catalase within cells is to prevent the accumulation of toxic levels of hydrogen peroxide formed as a byproduct of metabolic processes. Catalase might also take part in some of the many oxidation reactions that occur in all cells.

The primary reaction catalyzed by catalase is the decomposition of H_2O_2 to form water and oxygen:

$$\mathbf{2\ H_2O_2 \rightarrow 2\ H_2O + O_2\ (gas)}$$

In the absence of catalase, this reaction occurs spontaneously but very slowly. Catalase speeds up the reaction considerably. In this experiment, a rate for this reaction will be determined.

Much can be learned about enzymes by studying the kinetics (particularly the changes in rate) of enzyme-catalyzed reactions. For example, it is possible to measure the amount of product formed, or the amount of substrate used, from the moment the reactants are brought together until the reaction has stopped.

If the amount of product formed is measured at regular intervals and this quantity is plotted on a graph, a curve like the one in Figure 2.1 is obtained.

Figure 2.1: Enzyme Activity

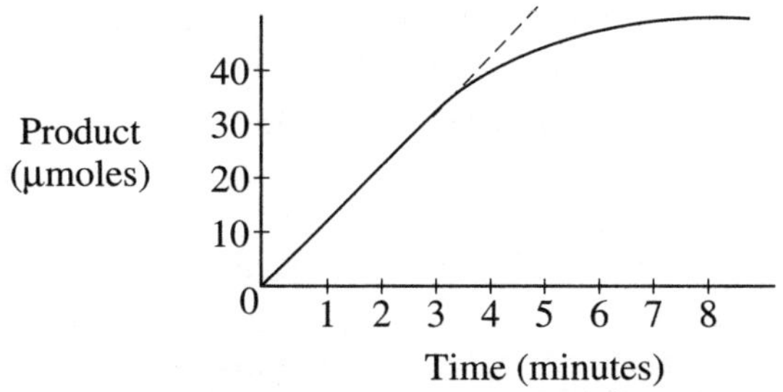

Study the solid line on the graph of this reaction. At time 0 there is no product. After 30 seconds, 5 micromoles (μmoles) have been formed; after one minute, 10 μmoles; after two minutes, 20 μmoles. The rate of this reaction could be given as 10 μmoles of product formed per minute for this initial period. Note, however, that by the third and fourth minutes, only about 5 additional μmoles of product have been formed. During the first three minutes, the rate is constant. From the third minute through the eighth minute, the rate is changing; it is slowing down. For each successive minute after the first three minutes, the amount of product formed in that interval is less than in the preceding minute. From the seventh minute onward, the reaction rate is very slow.

In the comparison of the kinetics of one reaction with another, a common reference point is needed. For example, suppose you wanted to compare the effectiveness of catalase obtained from potato with that of catalase obtained from liver. It is best to compare the reactions when the rates are constant. In the first few minutes of an enzymatic reaction such as this, the number of substrate molecules is usually so large compared with the number of enzyme molecules that changing the substrate concentration does not (for a short period at least) affect the number of successful collisions between substrate and enzyme. During this early period, the enzyme is acting on substrate molecules at a nearly constant rate. The slope of the graph line during this early period is called the **initial rate** of the reaction. The initial rate of any enzyme-catalyzed reaction is determined by the characteristics of the enzyme molecule. It is always the same for any enzyme and its substrate at a given temperature and pH. This also assumes that the substrate is present in excess.

The rate of the reaction is the slope of the linear portion of the curve. To determine a rate, pick any two points on the straight-line portion of the curve. Divide the difference in the amount of product formed between these two points by the difference in time between them. The result will be the rate of the reaction which, if properly calculated, can be expressed as μmoles product/sec. The rate, then, is:

$$\frac{\mu\text{moles}_2 - \mu\text{moles}_1}{t_2 - t_1}$$

or from the graph,

$$\frac{\Delta y}{\Delta x}$$

In the illustration of Figure 2.1, the rate between two and three minutes is calculated:

$$\frac{30 - 20}{180 - 120} = \frac{10}{60} = 0.17 \ \mu\text{moles/sec}$$

The rate of a chemical reaction may be studied in a number of ways, including the following:

1. measuring the rate of disappearance of substrate (in this example, H_2O_2);
2. measuring the rate of appearance of product (in this case, O_2, which is given off as a gas); and
3. measuring the heat released (or absorbed) during the reaction.

General Procedure

In this experiment the disappearance of the substrate, H_2O_2, is measured as follows (see Figure 2.2):

1. A purified catalase extract is mixed with substrate (H_2O_2) in a beaker. The enzyme catalyzes the conversion of H_2O_2 to H_2O and O_2 (gas).
2. Before all of the H_2O_2 is converted to H_2O and O_2, the reaction is stopped by adding sulfuric acid (H_2SO_4). The H_2SO_4 lowers the pH, denatures the enzyme, and thereby stops the enzyme's catalytic activity.
3. After the reaction is stopped, the amount of substrate (H_2O_2) remaining in the beaker is measured. To assay (measure) this quantity, potassium permanganate is used. Potassium permanganate ($KMnO_4$) in the presence of H_2O_2 and H_2SO_4 reacts as follows:

$$\mathbf{5\ H_2O_2 + 2\ KMnO_4 + 3\ H_2SO_4 \rightarrow K_2SO_4 + 2\ MnSO_4 + 8\ H_2O + 5\ O_2}$$

Note that H_2O_2 is a reactant for this reaction. Once all of the H_2O_2 has reacted, any more $KMnO_4$ added will be in excess and will not be decomposed. The addition of excess $KMnO_4$ causes the solution to have a permanent pink or brown color. Therefore, the amount of H_2O_2 remaining is determined by adding $KMnO_4$ until the whole mixture stays a faint pink or brown, permanently. Add no more $KMnO_4$ after this point. The amount of $KMnO_4$ added is a proportional measure of the amount of H_2O_2 remaining (2 molecules of $KMnO_4$ reacts with 5 molecules of H_2O_2 as shown in the equation).

Figure 2.2: The General Procedure

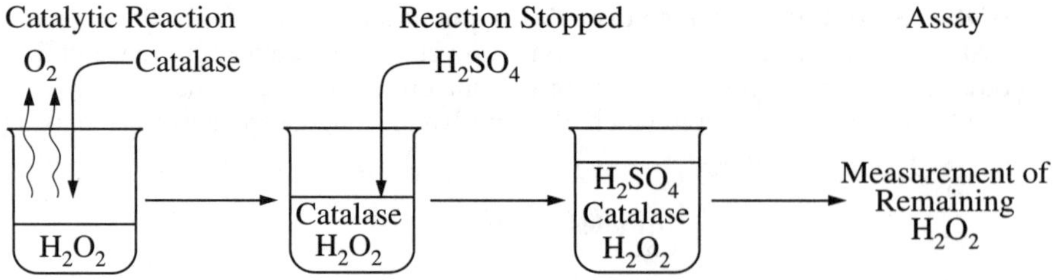

Figure 2.3: The Apparatus and Materials

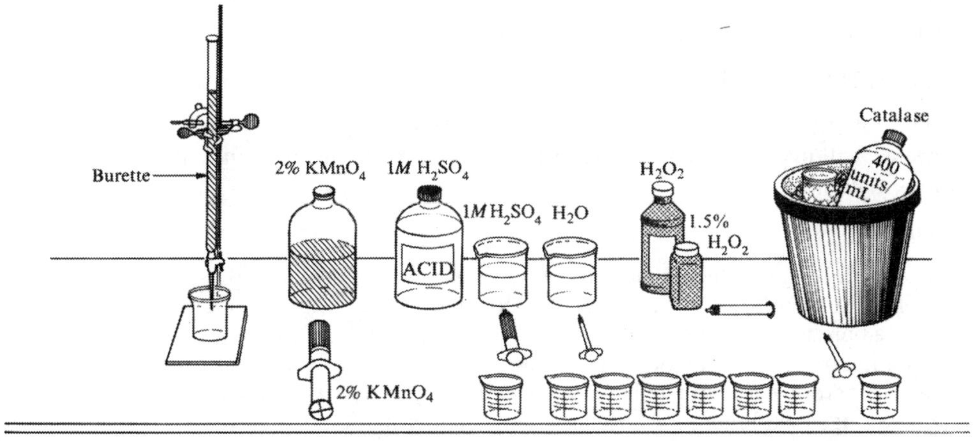

EXERCISE 2A: Test of Catalase Activity

Procedure

1. To observe the reaction to be studied, transfer 10 mL of 1.5% (0.44 *M*) H_2O_2 into a 50-mL glass beaker and add 1 mL of the freshly made catalase solution. The bubbles coming from the reaction mixture are O_2, which results from the breakdown of H_2O_2 by catalase. Be sure to keep the freshly made catalase solution on ice at all times.

   a. What is the enzyme in this reaction? ______________________

   b. What is the substrate in this reaction? ______________________

   c. What is the product in this reaction? ______________________

   d. How could you show that the gas evolved is O_2? ______________________

2. To demonstrate the effect of boiling on enzymatic activity, transfer 5 mL of purified catalase extract to a test tube and place it in a boiling water bath for five minutes. Transfer 10 mL of 1.5% H_2O_2 into a 50-mL glass beaker and add 1 mL of the cooled, boiled catalase solution. How does the reaction compare to the one using the unboiled catalase? Explain the reason for this difference.

   ____

   ____

3. To demonstrate the presence of catalase in living tissue, cut 1 cm^3 of potato or liver, macerate it, and transfer it into a 50-mL glass beaker containing 10 mL of 1.5% H_2O_2. What do you observe? What do you think would happen if the potato or liver was boiled before being added to the H_2O_2?

   ____

   ____

EXERCISE 2B: The Base Line Assay

To determine the amount of H_2O_2 initially present in a 1.5% solution, one needs to perform all the steps of the procedure without adding catalase (enzyme) to the reaction mixture. This amount is known as the base line and is an index of the initial concentration of H_2O_2 in solution. In any series of experiments, a base line should be established first.

Procedure for Establishing a Base Line

1. Put 10 mL of 1.5% H_2O_2 into a clean glass beaker.
2. Add 1 mL of H_2O (instead of enzyme solution).
3. Add 10 mL of H_2SO_4 (1.0 *M*). **Use extreme care in handling reagents.** Your teacher will instruct you about the proper safety procedures for handling hazardous materials.
4. Mix well.
5. *Remove a 5-mL sample.* Place this 5-mL sample into another beaker and assay for the amount of H_2O_2 as follows. Place the beaker containing the sample over a piece of white paper. Use a burette, a syringe, or a 5-mL pipette to add $KMnO_4$, a drop at a time, to the solution until a persistent pink or brown color is obtained. Remember to gently swirl the solution after adding each drop. Check to be sure that you understand the calibrations on the burette or syringe (see Figure 2.4). Record your reading in the box below.

Base line calculation

Final reading of burette _______ **mL**

Initial reading of burette _______ **mL**

Base line (Final – Initial) _______ **mL $KMnO_4$**

Figure 2.4: Proper Reading of a Burette

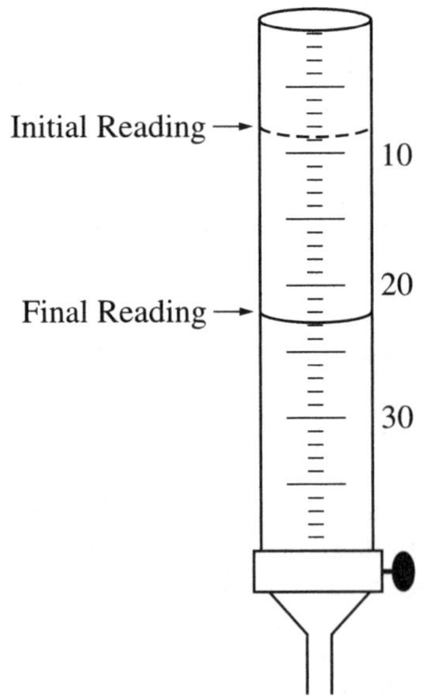

The base line assay value should be nearly the same for all groups. Compare your results to another team's before proceeding.

Remember, the amount of $KMnO_4$ used is proportional to the amount of H_2O_2 that was in the solution.

Note: Handle $KMnO_4$ with care. Avoid contact with skin and eyes.

EXERCISE 2C: The Uncatalyzed Rate of H_2O_2 Decomposition

To determine the rate of spontaneous conversion of H_2O_2 to H_2O and O_2 in an uncatalyzed reaction, put a small quantity of 1.5% H_2O_2 (about 15 mL) in a beaker. Store it uncovered at room temperature for approximately 24 hours. Repeat Steps 2–5 from Exercise 2B to determine the proportional amount of H_2O_2 remaining (for ease of calculation assume that 1 mL of $KMnO_4$ used in the titration represents the presence of 1 mL of H_2O_2 in the solution). Record your readings in the box below.

Uncatalyzed H_2O_2 decomposition

Final reading of burette __________ **mL**

Initial reading of burette __________ **mL**

Amount of $KMnO_4$ titrant __________ **mL**

Amount of H_2O_2 spontaneously decomposed (mL baseline – mL $KMnO_4$) __________ **mL**

What percent of the H_2O_2 spontaneously decomposes in 24 hours? [(mL baseline – mL 24 hours)/ mL baseline] × 100 __________ **%**

EXERCISE 2D: An Enzyme-Catalyzed Rate of H_2O_2 Decomposition

In this experiment you will determine the rate at which a 1.5% H_2O_2 solution decomposes when catalyzed by the purified catalase extract. To do this, you should determine how much H_2O_2 has been consumed after 10, 30, 60, 90, 120, 180, and 360 seconds.

If a day or more has passed since you did Exercise 2B, you must reestablish the base line by determining the amount of H_2O_2 present in your 1.5% solution. Repeat the assay procedure (Steps 1–5) and record your results in the box below. The base line assay should be approximately the same value for all groups. Check with another team before proceeding.

Base line calculation

Final reading of burette__________ **mL**

Initial reading of burette__________ **mL**

Base line (Final – Initial)__________ **mL $KMnO_4$**

Procedure for a Time-Course Determination

To determine the course of an enzymatic reaction, you will need to measure how much substrate is disappearing over time. You will measure the amount of substrate decomposed after 10, 30, 60, 90, 120, 180, and 360 seconds. To use lab time more efficiently, set up all of these at the same time and do them together. Stop each reaction at the proper time.

1. **10 seconds**
   - **a.** Put 10 mL of 1.5% H_2O_2 in a clean 50-mL glass beaker.
   - **b.** Add 1 mL of catalase extract.
   - **c.** Swirl gently for 10 seconds.
   - **d.** At 10 seconds, add 10 mL of H_2SO_4 (1.0 *M*).
2. **30, 60, 90, 120, 180, and 360 seconds**

   Each time, repeat Steps 1–4 as described above, except allow the reactions to proceed for 30, 60, 90, 120, 180, and 360 seconds, respectively, while swirling gently.

Note: Each time, remove a *5-mL* sample and assay for the amount of H_2O_2 in the sample. Use a burette to add $KMnO_4$, a drop at a time, to the solution until a persistent pink or brown color is obtained. Should the end point be overshot, remove another 5-mL sample and repeat the titration. *Do not discard any solutions until the entire lab is completed.* Record your results in Table 2.1 and Graph 2.1.

Table 2.1

$KMnO_4$ (ml)	Time (seconds)						
	10	30	60	90	120	180	360
a) Base Line*							
b) Final Reading							
c) Initial Reading							
d) Amount of $KMnO_4$ Consumed (B minus C)							
e) Amount of H_2O_2 Used (A minus D)							

3. Record the base line value, obtained in Exercise 2D, in all of the boxes on line A in Table 2.1.

* Remember that the base line tells how much H_2O_2 is in the initial 5-mL sample. The difference between the initial and final readings tells how much H_2O_2 is left after the enzyme-catalyzed reaction. The shorter the time, the more H_2O_2 remains and therefore the more $KMnO_4$ is necessary to titrate to the endpoint. If syringes are used, $KMnO_4$ consumed may be calculated as c−b.

4. Graph the data for enzyme-catalyzed H_2O_2 decomposition.

   For this graph you will need to determine the following:

   a. The *independent* variable: ________________________________
   Use this to label the horizontal (x) axis.

   b. The *dependent* variable: ________________________________
   Use this to label the vertical (y) axis.

Graph 2.1 Title: ____

Analysis of Results

1. From the formula described earlier recall that rate = $\frac{\Delta y}{\Delta x}$.

   Determine the initial rate of the reaction and the rates between each of the time points. Record the rates in the table below.

	Time Intervals (seconds)						
	Initial 0 to 10	10 to 30	30 to 60	60 to 90	90 to 120	120 to 180	180 to 360
Rates*							

2. When is the rate the highest? Explain why.

3. When is the rate the lowest? For what reasons is the rate low?

4. Explain the inhibiting effect of sulfuric acid on the function of catalase. Relate this to enzyme structure and chemistry.

5. Predict the effect that lowering the temperature would have on the rate of enzyme activity. Explain your prediction.

6. Design a controlled experiment to test the effect of varying pH, temperature, or enzyme concentration.

* Reaction rate (mL H_2O_2/sec)

LAB THREE MITOSIS AND MEIOSIS

OVERVIEW

In this lab you will investigate the processes of mitosis and meiosis:

1. You will use prepared slides of onion root tips to study plant mitosis and to calculate the relative duration of the phases of mitosis in the meristem of root tissue. Prepared slides of the whitefish blastula may be used to study mitosis in animal cells and to compare animal mitosis with plant mitosis.
2. You will simulate the stages of meiosis by using chromosome models. You will study the crossing over and recombination that occurs during meiosis. You will observe the arrangements of ascospores in the asci from a cross between wild type *Sordaria fimicola* and mutants for tan spore coat color in this fungus. These arrangements will be used to estimate the percentage of crossing over that occurs between the centromere and the gene that controls the tan spore color.

OBJECTIVES

Before doing this lab you should understand:

- the events of mitosis in animal and plant cells;
- the events of meiosis (gametogenesis in animals and sporogenesis in plants); and
- the key mechanical and genetic differences between meiosis and mitosis.

After doing this lab you should be able to:

- recognize the stages of mitosis in a plant or animal cell;
- calculate the relative duration of the cell cycle stages;
- describe how independent assortment and crossing over can generate genetic variation among the products of meiosis;
- use chromosome models to demonstrate the activity of chromosomes during meiosis I and meiosis II;
- relate chromosome activity to Mendel's laws of segregation and independent assortment;
- demonstrate the role of meiosis in the formation of gametes or spores in a controlled experiment using an organism of your choice;
- calculate the map distance of a particular gene from a chromosome's centromere or between two genes using an organism of your choice;
- compare and contrast the results of meiosis and mitosis in plant cells; and
- compare and contrast the results of meiosis and mitosis in animal cells.

INTRODUCTION

All new cells come from previously existing cells. New cells are formed by the process of cell division, which involves both division of the cell's nucleus **(karyokinesis)** and division of the cytoplasm **(cytokinesis).**

There are two types of nuclear division: mitosis and meiosis. **Mitosis** typically results in new somatic (body) cells. Formation of an adult organism from a fertilized egg, asexual reproduction, regeneration, and maintenance or repair of body parts are accomplished through mitotic cell division. You will study mitosis in Exercise 3A. **Meiosis** results in the formation of either gametes (in animals) or spores (in plants). These cells have half the chromosome number of the parent cell. You will study meiosis in Exercise 3B.

Where does one find cells undergoing mitosis? Plants and animals differ in this respect. In higher plants the process of forming new cells is restricted to special growth regions called **meristems.** These regions usually occur at the tips of stems or roots. In animals, cell division occurs anywhere new cells are formed or as new cells replace old ones. However, some tissues in both plants and animals rarely divide once the organism is mature.

To study the stages of mitosis, you need to look for tissues where there are many cells in the process of mitosis. This restricts your search to the tips of growing plants, such as the onion root tip, or, in the case of animals, to developing embryos, such as the whitefish blastula.

EXERCISE 3A.1: Observing Mitosis in Plant and Animal Cells Using Prepared Slides of the Onion Root Tip and Whitefish Blastula

Roots consist of different regions (see Figure 3.1a). The **root cap** functions in protection. The **apical meristem** (Figure 3.1b) is the region that contains the highest percentage of cells undergoing mitosis. The **region of elongation** is the area in which growth occurs. The **region of maturation** is where root hairs develop and where cells differentiate to become xylem, phloem, and other tissues.

Figure 3.1a: Median Longitudinal Section

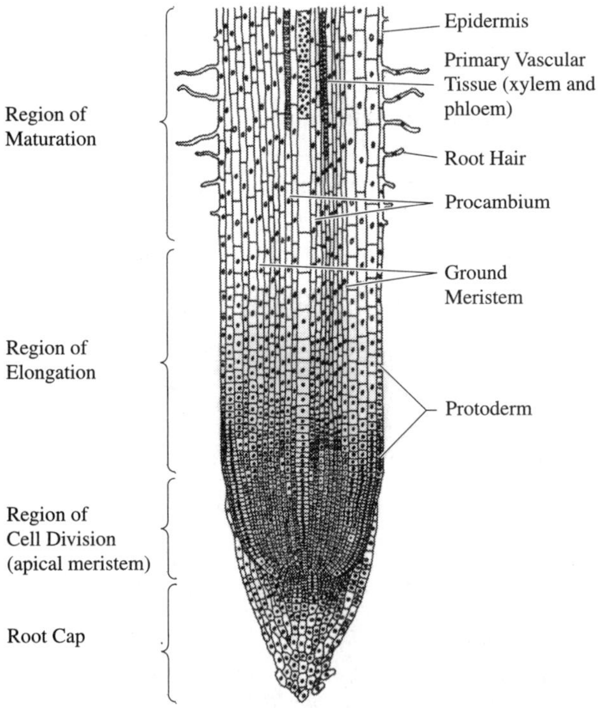

Figure 3.1b: Apical Meristem Tip Close Up

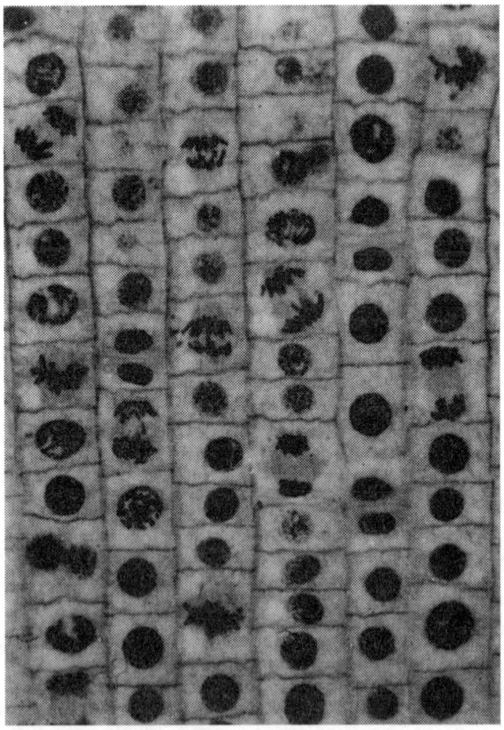

Figure 3.2: Whitefish Blastula

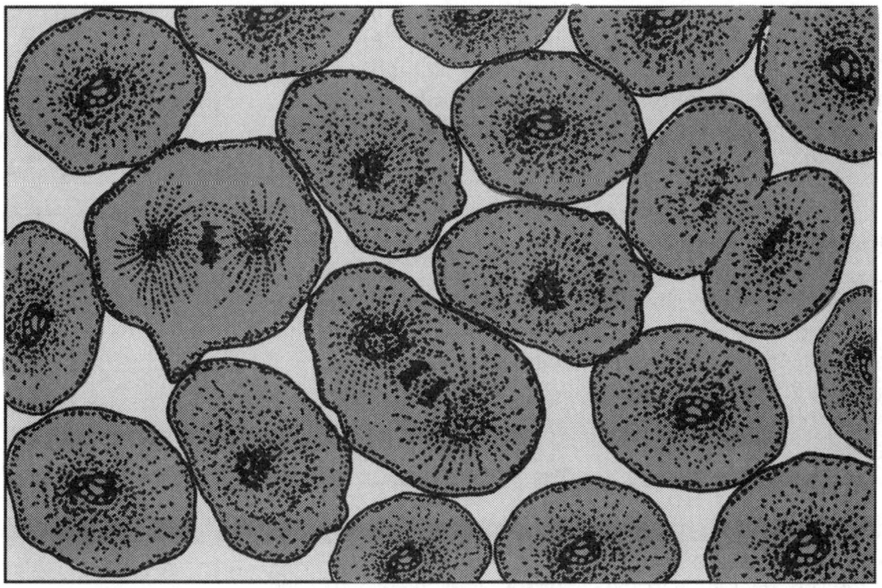

The whitefish blastula is often used for the study of cell division. As soon as the egg is fertilized it begins to divide, and nuclear division after nuclear division follows. You will be provided with slides of whitefish blastula which have been sectioned in various planes in relation to the mitotic spindle. You will be able to see side and polar views of the spindle apparatus.

Procedure

Examine prepared slides of either onion root tips or whitefish blastula. Locate the meristematic region of the onion, or locate the blastula, with the 10X objective and then use the 40X objective to study individual cells. For convenience in discussion, biologists have described certain stages, or phases, of the continuous mitotic cell cycle, as outlined on this page and the next. Identify one cell that clearly represents each phase. Sketch and label the cell in the boxes provided.

---

1. The nondividing cell is in a stage called **interphase.** The **nucleus** may have one or more dark-stained **nucleoli** and is filled with a fine network of threads, the **chromatin.** During **interphase,** DNA replication occurs.

Interphase

Figure 3.3

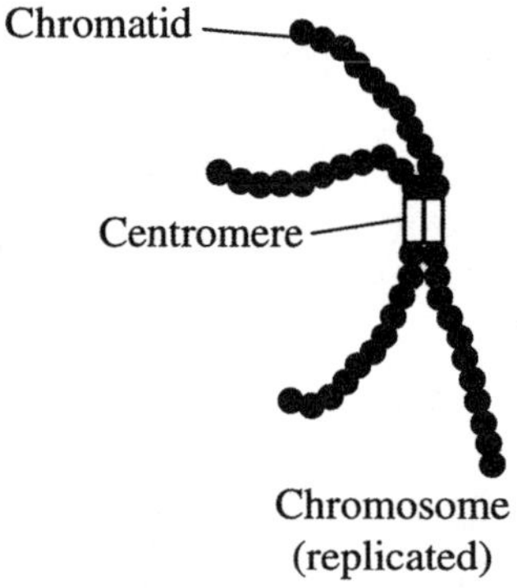

2. The first sign of division occurs in **prophase.** There is a thickening of the chromatin threads, which continues until it is evident that the chromatin has condensed into **chromosomes** (Figure 3.3). With somewhat higher magnification you may be able to see that each chromosome is composed of two **chromatids** joined at a **centromere.** As prophase continues, the chromatids continue to shorten and thicken. In late prophase the nuclear envelope and nucleoli are no longer visible, and the chromosomes are free in the cytoplasm. Just before this time, the first sign of a spindle appears in the cytoplasm; the spindle apparatus is made up of **microtubules,** and it is thought that these microtubules may pull the chromosomes toward the poles of the spindle where the two daughter nuclei will eventually form.

Prophase

---

3. At **metaphase** the chromosomes have moved to the center of the spindle. One particular portion of each chromosome, the centromere, attaches to the spindle. The centromeres of all the chromosomes lie at about the same level of the spindle, on a plane called the metaphase plate. At metaphase you should be able to observe the two chromatids of some of the chromosomes.

Metaphase

---

4. At the beginning of **anaphase,** the centromere regions of each pair of chromatids separate and are moved by the spindle fibers toward opposite poles of the spindle, dragging the rest of the chromatid behind them. Once the two chromatids separate, each is called a **chromosome**. These daughter chromosomes continue their poleward movement until they form two compact clumps, one at each spindle pole.

Anaphase

---

5. **Telophase,** the last stage of division, is marked by a pronounced condensation of the chromosomes, followed by the formation of a new nuclear envelope around each group of chromosomes. The chromosomes gradually uncoil to form the fine chromatin network seen in interphase, and the nucleoli and nuclear envelope reappear. **Cytokinesis** may occur. This is the division of the cytoplasm into two cells. In plants, a new cell wall is laid down between the daughter cells. In animal cells, the old cell will pinch off in the middle along a **cleavage furrow** to form two new daughter cells.

Telophase

Analysis Questions

1. Explain how mitosis leads to two daughter cells, each of which is diploid and genetically identical to the original cell. What activities are going on in the cell during interphase?

2. How does mitosis differ in plant and animal cells? How does plant mitosis accommodate a rigid, inflexible cell wall?

3. What is the role of the centrosome (the area surrounding the centrioles)? Is it necessary for mitosis? Defend your answer.

EXERCISE 3A.2: Time for Cell Replication

To estimate the relative length of time that a cell spends in the various stages of cell division, you will examine the meristematic region of a prepared slide of the onion root tip. The length of the cell cycle is approximately 24 hours for cells in actively dividing onion root tips.

Procedure

It is hard to imagine that you can estimate how much time a cell spends in each phase of cell division from a slide of dead cells, yet this is precisely what you will do in this part of the lab. Since you are working with a prepared slide, you cannot get any information about how long it takes a cell to divide. What you can determine is how many cells are in each phase. From this, you can infer the percentage of time each cell spends in each phase.

1. Observe every cell in one high-power field of view and determine which phase of the cell cycle it is in. This is best done in pairs. The partner observing the slide calls out the phase of each cell while the other partner records. Then switch so the recorder becomes the observer and vice versa. Count at least two full fields of view. If you have not counted at least 200 cells, then count a third field of view.

2. Record your data in Table 3.1.

3. Calculate the percentage of cells in each phase, and record in Table 3.1.

Consider that it takes, on average, 24 hours (or 1,440 minutes) for onion root tip cells to complete the cell cycle. You can calculate the amount of time spent in each phase of the cell cycle from the percentage of cells in that stage.

Percentage of cells in stage × 1,440 minutes = ________ minutes of cell cycle spent in stage

Table 3.1

	Number of Cells				Percent of Total Cells Counted	Time in Each Stage
	Field 1	Field 2	Field 3	Total		
Interphase						
Prophase						
Metaphase						
Anaphase						
Telophase						
			Total Cells Counted			

Questions

1. If your observations had not been restricted to the area of the root tip that is actively dividing, how would your results have been different?

____

____

____

____

2. Based on the data in Table 3.1, what can you infer about the relative length of time an onion root tip cell spends in each stage of cell division?

____

____

____

____

3. Draw and label a pie chart of the onion root tip cell cycle using the data from Table 3.1.

Title: ____________

EXERCISE 3B: Meiosis

Meiosis involves two successive nuclear divisions that produce four haploid cells. **Meiosis I** is the reduction division. It is this first division that reduces the chromosome number from diploid to haploid and separates the homologous pairs. **Meiosis II,** the second division, separates the sister chromatids. The result is four haploid gametes.

Mitotic cell division produces new cells genetically identical to the parent cell. Meiosis increases genetic variation in the population. Each diploid cell undergoing meiosis can produce 2^n different chromosomal combinations, where n is the haploid number. In humans the number is 2^{23}, which is more than eight million different combinations. Actually, the potential variation is even greater because, during meiosis I, each pair of chromosomes (homologous chromosomes) comes together in a process known as **synapsis.** Chromatids of homologous chromosomes may exchange parts in a process called **crossing over.** The relative distance between two genes on a given chromosome can be estimated by calculating the percentage of crossing over that takes place between them.

EXERCISE 3B.1: Simulation of Meiosis

In this exercise you will study the process of meiosis by using chromosome simulation kits and following the directions in Figures 3.4 –3.8. Your kit should contain two strands of beads of one color and two strands of another color. A homologous pair of chromosomes is represented by *one* strand of each color, with one of each pair coming from each parent. The second strands of each color are to be used as chromatids for each of these chromosomes.

Figure 3.4

Interphase. Place one strand of each color near the center of your work area. (Recall that chromosomes at this stage would exist as diffuse chromatin and not as visible structures.) DNA synthesis occurs during interphase, and each chromosome, originally composed of one strand, is now made up of two strands, or chromatids, joined together at the centromere region. Simulate DNA replication by bringing the magnetic centromere region of one strand in contact with the centromere region of the other of the same color. Do the same with its homolog.

Summary: DNA replication

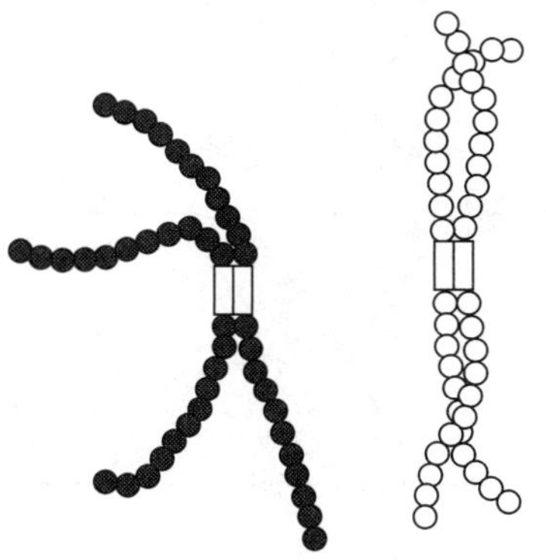

Figure 3.5

Prophase I. Homologous chromosomes come together and synapse along their entire length. **This pairing, or synapsis, of homologous chromosomes represents the first big difference between mitosis and meiosis.** A **tetrad,** consisting of four chromatids, is formed. Use the models of two chromosomes to simulate synapsis and the process of crossing over. Crossing over can be simulated by popping the beads apart on one chromatid at the fifth bead, or "gene," and doing the same with the other chromatid. Reconnect the beads to those of the other color. Proceed through prophase I of meiosis and note how crossing over results in recombination of genetic information. The visual result of crossing over is called a **chiasma** (plural **chiasmata**).

Summary: Synapsis and crossing over

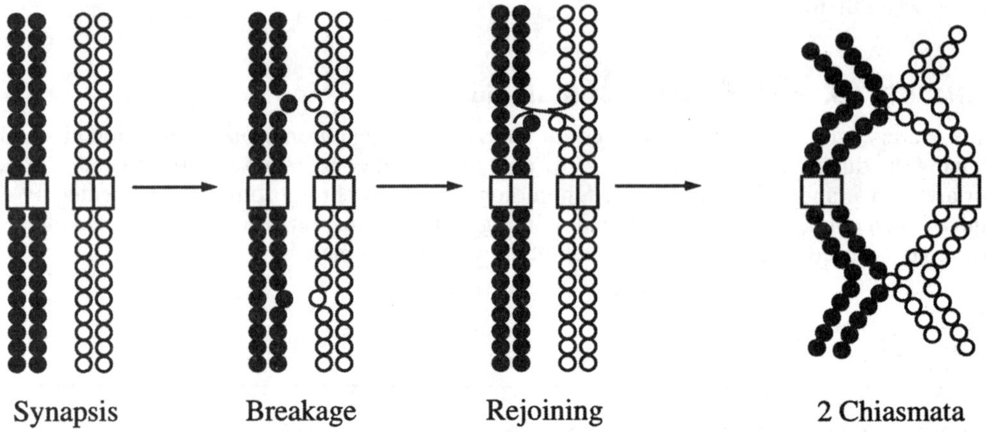

Figure 3.6

Metaphase I. The crossed-over tetrads line up in the center of the cell. Position the chromosomes near the middle of the cell.

Summary: Tetrads align on equator

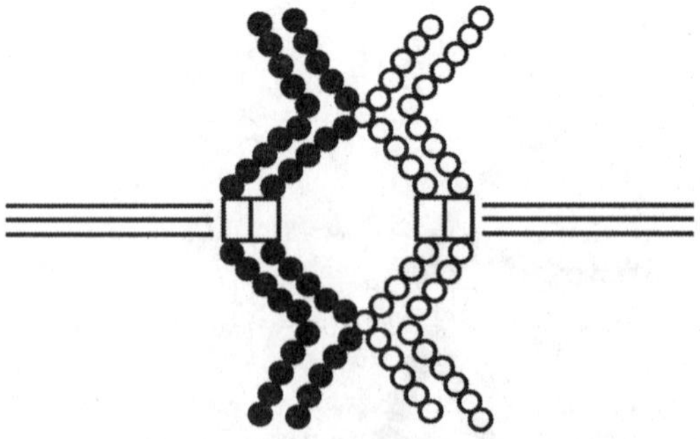

Figure 3.7

Anaphase I. During anaphase I the homologous chromosomes separate and are "pulled" to opposite sides of the cell. **This represents a second significant difference between the events of mitosis and meiosis.**

Summary: Homologs separate
Chromosome number is reduced

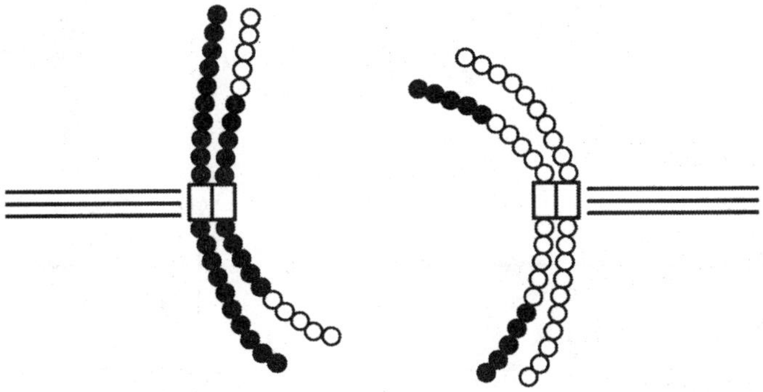

Figure 3.8

Telophase I. Place each chromosome at opposite sides of the cell. Formation of a nuclear envelope and division of the cytoplasm (cytokinesis) often occur at this time to produce two cells, but this is not always the case. Notice that each chromosome within the two daughter cells still consists of two chromatids.

Summary: 2 haploid cells formed
Each chromosome composed of 2 chromatids

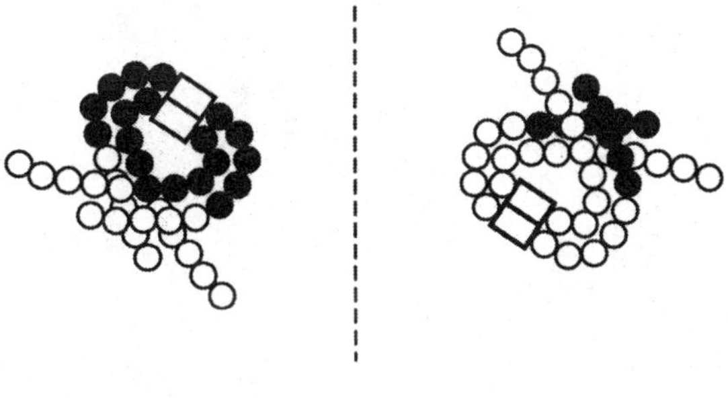

Interphase II (Interkinesis). The amount of time spent "at rest" following telophase I depends on the type of organism, the formation of new nuclear envelopes, and the degree of chromosomal uncoiling. Because interphase II does not necessarily resemble interphase I, it is often given a different name — **interkinesis. DNA replication does not occur during interkinesis. This represents a third difference between mitosis and meiosis.**

Meiosis II

A second meiotic division is necessary to separate the chromatids of the chromosomes in the two daughter cells formed by this first division. This will reduce the amount of DNA to one strand per chromosome. This second division is called meiosis II. It resembles mitosis except that only one homolog from each homologous pair of chromosomes is present in each daughter cell undergoing meiosis II.

The following simulation procedures apply to haploid nuclei produced by meiosis I.

Figure 3.9

Prophase II. No DNA replication occurs. Replicated centrioles (not shown) separate and move to opposite sides of the chromosome groups.

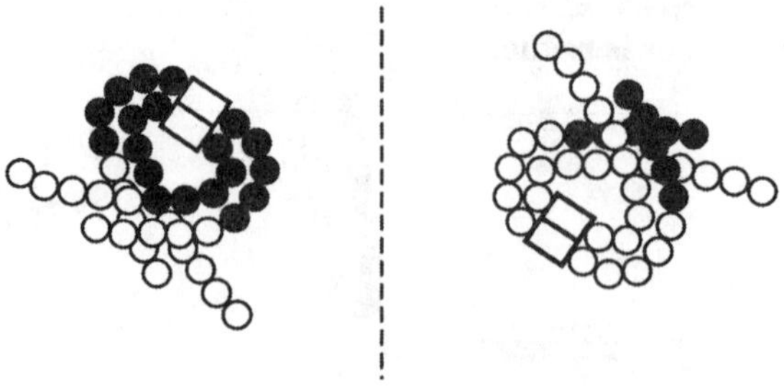

Figure 3.10

Metaphase II. Orient the chromosomes so that they are centered in the middle of each daughter cell.

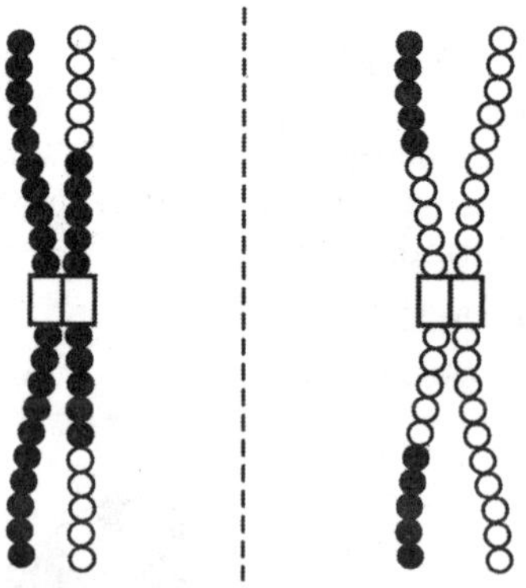

Figure 3.11

Anaphase II. The centromere regions of the chromatids now appear to be separate. Separate the chromatids of the chromosomes and pull the daughter chromosomes toward the opposite sides of each daughter cell. Now that each chromatid has its own visibly separate centromere region, it can be called a chromosome.

Summary: Chromatids separate

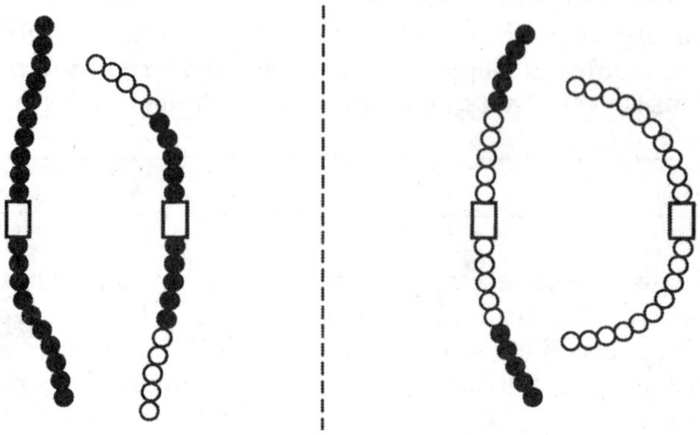

Figure 3.12

Telophase II. Place the chromosomes at opposite sides of the dividing cell. At this time a nuclear envelope forms and, in our simulation, the cytoplasm divides.

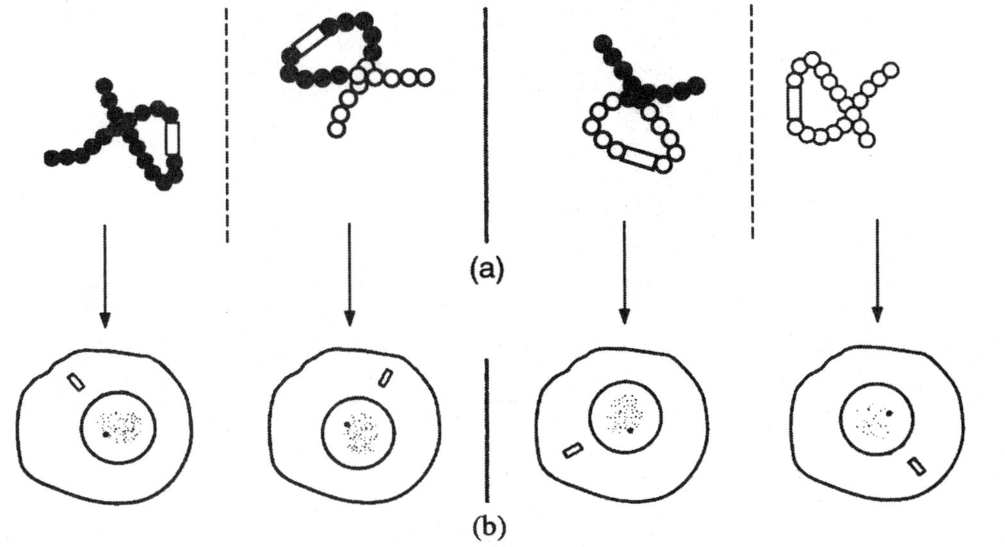

Analysis and Investigation

1. List three major differences between the events of mitosis and meiosis.

2. Compare mitosis and meiosis with respect to each of the following in Table 3.2:

Table 3.2

	Mitosis	Meiosis
Chromosome Number of Parent Cells		
Number of DNA Replications		
Number of Divisions		
Number of Daughter Cells Produced		
Chromosome Number of Daughter Cells		
Purpose/Function		

3. How are meiosis I and meiosis II different?

4. How do oogenesis and spermatogenesis differ?

5. Why is meiosis important for sexual reproduction?

EXERCISE 3B.2: Crossing Over during Meiosis in *Sordaria*

Sordaria fimicola is an ascomycete fungus that can be used to demonstrate the results of crossing over during meiosis. *Sordaria* is a **haploid** organism for most of its life cycle. It becomes **diploid** only when the fusion of the mycelia (filamentlike groups of cells) of two different strains results in the fusion of the two different types of haploid nuclei to form a diploid nucleus. The diploid nucleus must then undergo meiosis to resume its haploid state.

Meiosis, followed by one mitotic division, in *Sordaria* results in the formation of eight **haploid ascospores** contained within a sac called an **ascus** (plural, **asci**). Many asci are contained within a fruiting body called a **perithecium** (ascocarp). When ascospores are mature the ascus ruptures, releasing the ascospores. Each ascospore can develop into a new haploid fungus. The life cycle of *Sordaria fimicola* is shown in Figure 3.13.

Figure 3.13: The Life Cycle of *Sordaria fimicola*

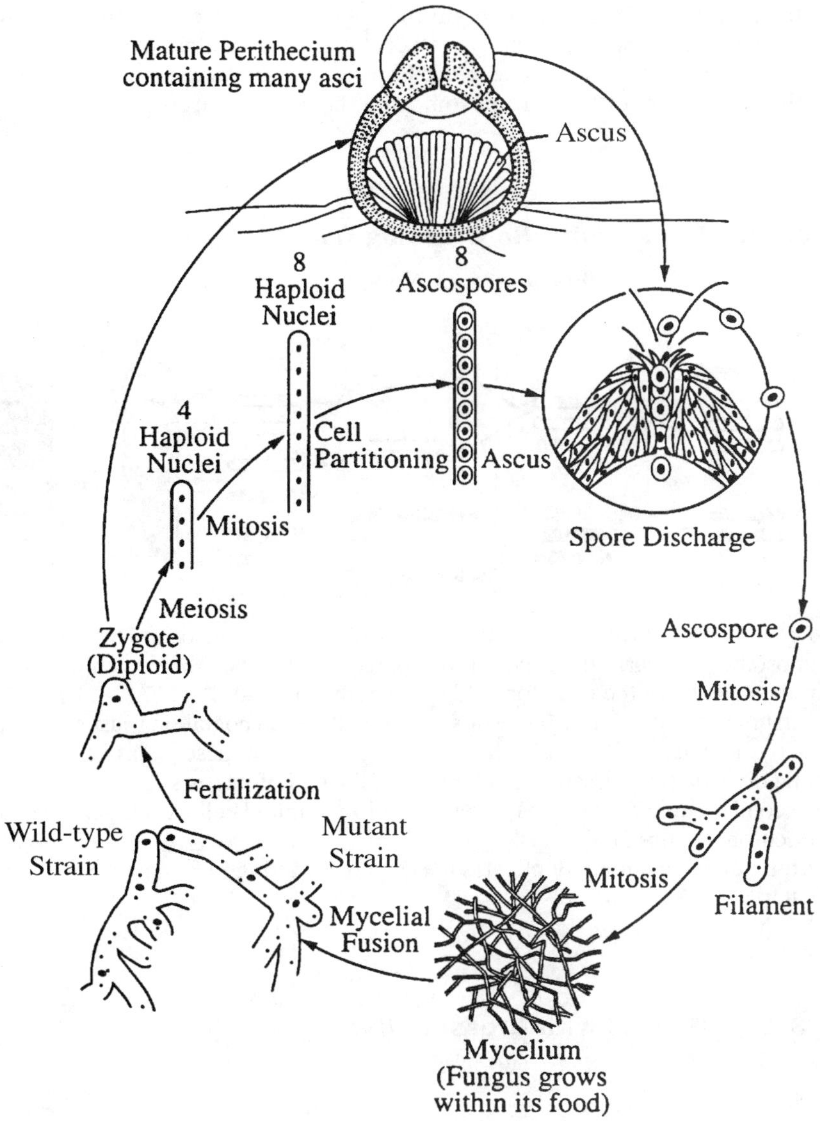

To observe crossing over in *Sordaria*, one must make hybrids between wild type and mutant strains of *Sordaria*. Wild type *Sordaria* have black ascospores (+). One mutant strain has tan spores (tn). When mycelia of these two different strains come together and undergo meiosis, the asci that develop will contain four black ascospores and four tan ascospores. The arrangement of the spores directly reflects whether or not crossing over has occurred. In Figure 3.14 no crossing over has occurred. Figure 3.15 shows the results of crossing over between the centromere of the chromosome and the gene for ascospore color.

Figure 3.14: Meiosis with No Crossing Over

Formation of Noncrossover Asci

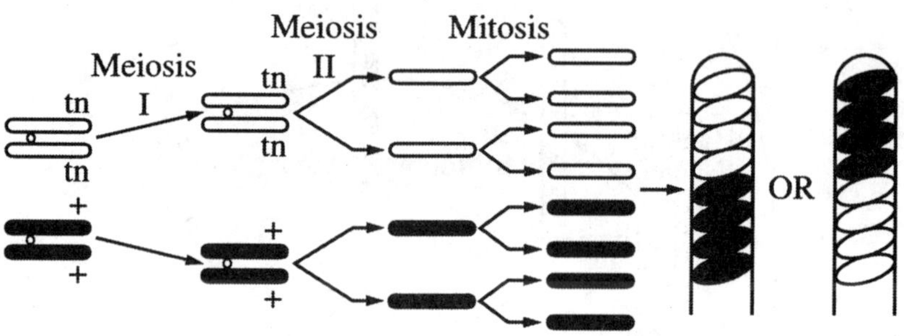

Two homologous chromosomes line up at metaphase I of meiosis. The two chromatids of one chromosome each carry the gene for tan spore color (tn) and the two chromatids of the other chromosome carry the gene for wild type spore color (+).

The first meiotic division (MI) results in two cells, each containing just one type of spore color gene (either tan or wild type). Therefore, segregation of these genes has occurred at the first meiotic division (MI). Each cell is haploid at the end of meiosis I.

The second meiotic division (MII) results in four haploid cells, each with the haploid number of chromosomes (1N).

A **mitotic** division simply duplicates these cells, resulting in 8 spores. They are arranged in the 4:4 pattern.

Figure 3.15: Meiosis with Crossing Over

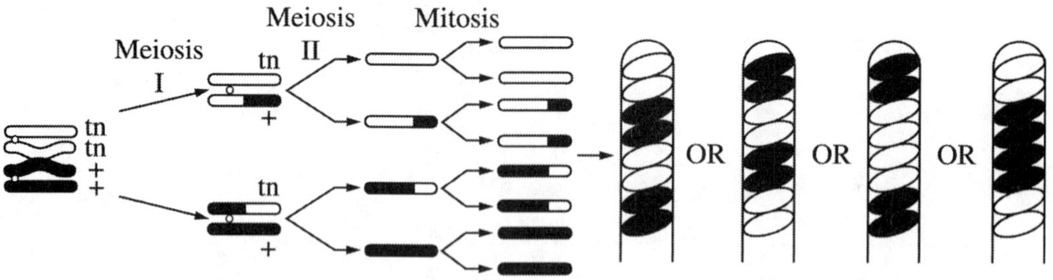

In this example crossing over has occurred in the region between the gene for spore color and the centromere. The homologous chromosomes separate during meiosis I.

This time the MI results in two cells, each containing both genes (1 tan, 1 wild type); therefore, the genes for spore color have not yet segregated, although the cells are haploid.

Meiosis II (MII) results in segregation of the two types of genes for spore color.

A **mitotic** division results in 8 spores arranged in the 2:2:2:2 or 2:4:2 pattern. Any one of these spore arrangements would indicate that crossing over has occurred between the gene for spore coat color and the centromere.

Procedure

1. Two strains of *Sordaria* (wild type and tan mutant) have been inoculated on a plate of agar. Where the mycelia of the two strains meet (Figure 3.16), fruiting bodies called perithecia develop. Meiosis occurs within the perithecia during the formation of asci. Use a toothpick to gently scrape the surface of the agar to collect perithecia (the black dots in the figure below).

Figure 3.16

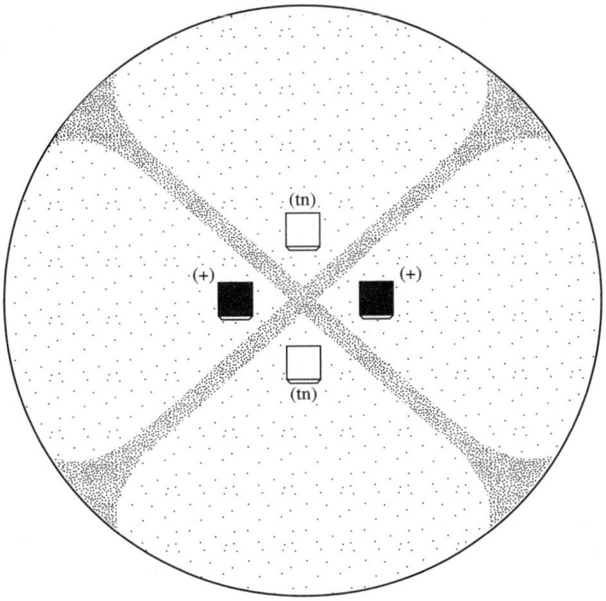

2. Place the perithecia in a drop of water or glycerin on a slide. Cover with a cover slip and return to your workbench. Using the eraser end of a pencil, press the cover slip down gently so that the perithecia rupture but the ascospores remain in the asci. Using the 10X objective, view the slide and locate a group of hybrid asci (those containing both tan and black ascospores). Count at least 50 hybrid asci and enter your data in Table 3.3.

Table 3.3

Number of 4:4 ●●●●○○○○ ○○○○●●●●	Number of Asci Showing Crossover ○○●●○○●● ●●○○●●○○ ○○●●●●○○ ●●○○○○●●	Total Asci	% Asci Showing Crossover Divided by 2	Gene to Centromere Distance (map units)

The frequency of crossing over appears to be governed largely by the distance between genes, or in this case, between the gene for spore coat color and the centromere. The probability of a crossover occurring between two particular genes on the same chromosome (linked genes) increases as the distance between those genes becomes larger. The frequency of crossover, therefore, appears to be directly proportional to the distance between genes.

A **map unit** is an arbitrary unit of measure used to describe relative distances between linked genes. The number of map units between two genes or between a gene and the centromere is equal to the percentage of recombinants. Customary units cannot be used because we cannot directly visualize genes with the light microscope. However, due to the relationship between distance and crossover frequency, we may use the map unit.

Analysis of Results

1. Using your data in Table 3.3, determine the distance between the gene for spore color and the centromere. Calculate the percentage of crossovers by dividing the number of crossover asci (2:2:2:2 or 2:4:2) by the total number of asci × 100. To calculate the map distance, divide the percentage of crossover asci by 2. The percentage of crossover asci is divided by 2 because only half of the spores in each ascus are the result of a crossover event (Figure 3.15). Record your results in Table 3.3.

2. Draw a pair of chromosomes in MI and MII and show how you would get a 2:4:2 arrangement of ascospores by crossing over. (Hint: refer to Figure 3.15).

LAB FOUR PLANT PIGMENTS AND PHOTOSYNTHESIS

OVERVIEW

In this lab you will:

1. separate plant pigments using chromatography, and
2. measure the rate of photosynthesis in isolated chloroplasts using the dye DPIP.

The transfer of electrons during the light-dependent reactions of photosynthesis reduces DPIP, changing it from blue to colorless.

OBJECTIVES

Before doing this lab you should understand:

- how chromatography separates two or more compounds that are initially present in a mixture;
- the process of photosynthesis;
- the function of plant pigments;
- the relationship between light wavelength and photosynthetic rate; and
- the relationship between light intensity and photosynthetic rate.

After doing this lab you should be able to:

- separate pigments and calculate their R_f values;
- describe a technique to determine photosynthetic rates;
- compare photosynthetic rates at different light intensities or different wavelengths of light using controlled experiments; and
- explain why the rate of photosynthesis varies under different environmental conditions.

EXERCISE 4A: Plant Pigment Chromatography

Paper chromatography is a useful technique for separating and identifying pigments and other molecules from cell extracts that contain a complex mixture of molecules. The solvent moves up the paper by capillary action, which occurs as a result of the attraction of solvent molecules to the paper and the attraction of solvent molecules to one another. As the solvent moves up the paper, it carries along any substances dissolved in it. The pigments are carried along at different rates because they are not equally soluble in the solvent and because they are attracted, to different degrees, to the fibers in the paper through the formation of intermolecular bonds, such as hydrogen bonds.

Beta carotene, the most abundant carotene in plants, is carried along near the solvent front because it is very soluble in the solvent being used and because it forms no hydrogen bonds with cellulose. Another pigment, xanthophyll, differs from carotene in that it contains oxygen. Xanthophyll is found further from the solvent front because it is less soluble in the solvent and has been slowed down by hydrogen bonding to the cellulose. Chlorophylls contain oxygen and nitrogen and are bound more tightly to the paper than are the other pigments.

Chlorophyll *a* is the primary photosynthetic pigment in plants. A molecule of chlorophyll *a* is located at the reaction center of photosystems. Other chlorophyll *a* molecules, chlorophyll *b*, and the carotenoids (that is, carotenes and xanthophylls) capture light energy and transfer it to the chlorophyll *a* at the reaction center. Carotenoids also protect the photosynthetic system from the damaging effects of ultraviolet light.

Procedure

Your teacher will demonstrate the apparatus and techniques used in paper chromatography. Here is a suggested procedure, illustrated in Figure 4.1:

1. Obtain a 50-ml graduated cylinder that has 1 cm of solvent in the bottom. The cylinder is tightly stoppered because this solvent is volatile, and you should be careful to keep the stopper on as much as possible.
2. Cut a piece of filter paper that will be long enough to reach the solvent. Cut one end ofthis filter paper into a point. Draw a pencil line 1.5 cm above the point.
3. Use a coin to extract the pigments from spinach leaf cells. Place a small section of leaf on the top of the pencil line. Use the ribbed edge of the coin to crush the cells. Be sure that the pigment line is on top of the pencil line. You should repeat this procedure 8 to 10 times, being sure to use a new portion of the leaf each time.
4. Place the chromatography paper in the cylinder so that the pointed end is barely immersed in the solvent. *Do not allow the pigment to be in the solvent.*
5. Stopper the cylinder. When the solvent is about 1 cm from the top of the paper, remove the paper and *immediately* mark the location of the solvent front before it evaporates.
6. Mark the bottom of each pigment band. Measure the distance each pigment migrated from the bottom of the pigment origin to the bottom of the separated pigment band. In Table 4.1 record the distance that each front, including the solvent front, moved. Depending on the species of plant used, you may be able to observe 4 or 5 pigment bands.

Figure 4.1

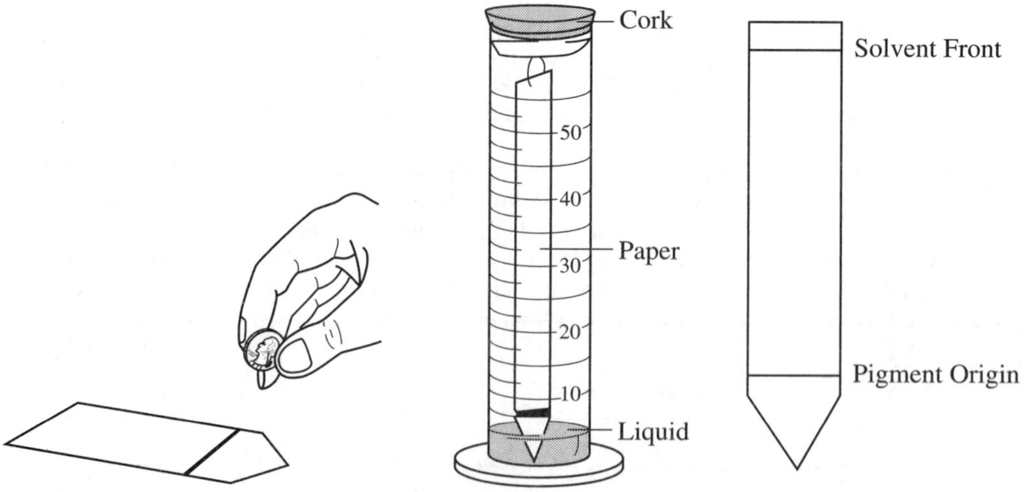

Table 4.1

Distance Moved by Pigment Band (millimeters)

Band Number	Distance (mm)	Band Color
1.		
2.		
3.		
4.		
5.		

Distance Solvent Front Moved _________ (mm)

Analysis of Results

The relationship of the distance moved by a pigment to the distance moved by the solvent is a constant called $\mathbf{R_f}$. It can be calculated for each of the four pigments using the following formula:

$$R_f = \frac{\text{distance pigment migrated (mm)}}{\text{distance solvent front migrated (mm)}}$$

Record your $\mathbf{R_f}$ values in Table 4.2.

Table 4.2

_______________ = R_f for Carotene (yellow to yellow orange)

_______________ = R_f for Xanthophyll (yellow)

_______________ = R_f for Chlorophyll *a* (bright green to blue green)

_______________ = R_f for Chlorophyll *b* (yellow green to olive green)

Topics for Discussion

1. What factors are involved in the separation of the pigments?

____

____

____

____

____

____

2. Would you expect the R_f value of a pigment to be the same if a different solvent were used? Explain.

3. What type of chlorophyll does the reaction center contain? What are the roles of the other pigments?

EXERCISE 4B: Photosynthesis/The Light Reaction

Light is a part of a continuum of radiation, or energy waves. Shorter wavelengths of energy have greater amounts of energy. For example, high-energy ultraviolet rays can harm living tissues. Wavelengths of light within the visible part of the light spectrum power photosynthesis.

When light is absorbed by leaf pigments, electrons within each photosystem are boosted to a higher energy level, and this energy is used to produce ATP and to reduce NADP to NADPH. ATP and NADPH are then used to incorporate CO_2 into organic molecules, a process called **carbon fixation.**

Design of the Exercise

Photosynthesis may be studied in a number of ways. For this experiment a dye-reduction technique will be used. The dye-reduction experiment tests the hypothesis that light and chloroplasts are required for the light reactions to occur. In place of the electron acceptor, NADP, the compound DPIP (2,6-dichlorophenol-indophenol), will be substituted. When light strikes the chloroplasts, electrons boosted to high energy levels will reduce DPIP. It will change from blue to colorless.

In this experiment chloroplasts are extracted from spinach leaves and incubated with DPIP in the presence of light. As the DPIP is reduced and becomes colorless, the resultant increase in light transmittance is measured over a period of time using a spectrophotometer. The experimental design matrix is presented in Table 4.3.

Table 4.3: Photosynthesis Setup

	Cuvettes				
	1 Blank (no DPIP)	2 Unboiled Chloroplasts Dark	3 Unboiled Chloroplasts Light	4 Boiled Chloroplasts Light	5 No Chloroplasts Light
Phosphate Buffer	1 mL	1 mL	1 mL	1 mL	1 mL
Distilled H_2O	4 mL	3 mL	3 mL	3 mL	3 mL + 3 drops
DPIP	—	1 mL	1 mL	1 mL	1 mL
Unboiled Chloroplasts	3 drops	3 drops	3 drops	—	—
Boiled Chloroplasts	—	—	—	3 drops	—

Procedure

1. Turn on the spectrophotometer to warm up the instrument and set the wavelength to 605 nm by adjusting the wavelength control knob.
2. While the spectrophotometer is warming up, your teacher may demonstrate how to prepare a chloroplast suspension from spinach leaves.
3. Set up an incubation area that includes a light, water flask, and test tube rack (see Figure 4.2). The water in the flask acts as a heat sink by absorbing most of the light's infrared radiation while having little effect on the light's visible radiation.

Figure 4.2: Incubation Setup

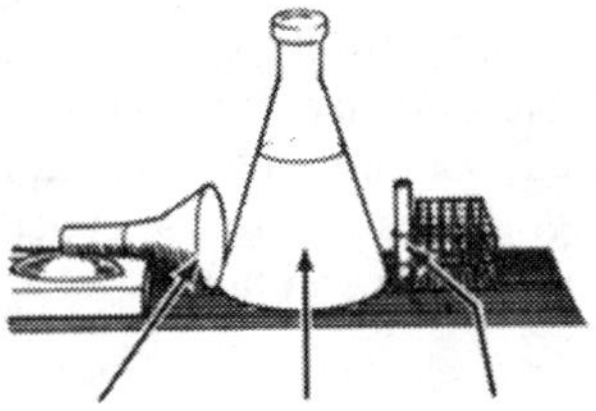

4. Your teacher will provide you with two beakers, one containing a solution of boiled chloroplasts and the other one containing unboiled chloroplasts. Be sure to keep both beakers on ice at all times.
5. At the top rim, label the cuvettes 1, 2, 3, 4, and 5, respectively. Be sure to follow your teacher's directions on how to label cuvettes. Using lens tissue, wipe the outside walls of each cuvette *(Remember: handle cuvettes only near the top).* Cover the walls and bottom of cuvette 2 with foil and make a foil cap to cover the top. Light should not be permitted inside cuvette 2 because it is a control for this experiment.
6. Refer to Table 4.3 to prepare each cuvette. *Do not add unboiled or boiled chloroplasts yet.* To each cuvette, add 1 mL of phosphate buffer. To cuvette 1, add 4 mL of distilled H_2O. To cuvettes 2, 3, and 4, add 3 mL of distilled H_2O and 1 mL of DPIP. To cuvette 5, add 3 mL plus 3 drops of distilled water, and 1 mL of DPIP.

7. Bring the spectrophotometer to zero by adjusting the amplifier control knob until the meter reads 0% transmittance. Add 3 drops of unboiled chloroplasts to cuvette 1. Cover the top with Parafilm® and invert to mix. Insert cuvette 1 into the sample holder and adjust the instrument to 100% transmittance by adjusting the light-control knob. *Cuvette 1 is the blank to be used to recalibrate the instrument between readings.* In other words, you will measure the light transmitted through each of the other tubes as a percentage of the light transmitted through this tube. For each reading, make sure that the cuvettes are inserted into the sample holder so that they face the same way as in the previous reading.

8. Obtain the unboiled chloroplast suspension, stir to mix, and transfer 3 drops to cuvette 2. *Immediately* cover and mix cuvette 2. Then remove it from the foil sleeve and insert it into the spectrophotometer's sample holder, read the % transmittance, and record it as the time 0 reading in Table 4.4. Replace cuvette 2 in the foil sleeve and place it in the incubation test tube rack. Turn on the flood light. Take and record additional readings at 5, 10, and 15 minutes. Mix the cuvette's contents just prior to each reading. Remember to use cuvette 1 occasionally to check and adjust the spectrophotometer to 100% transmittance.

9. Obtain the unboiled chloroplast suspension, mix, and transfer 3 drops to cuvette 3. *Immediately* cover and mix cuvette 3. Insert into the sample holder, read the % transmittance, and record it in Table 4.4. Place cuvette 3 in the incubation test tube rack next to cuvette 2. Take and record additional readings at 5, 10, and 15 minutes. Mix the cuvette's contents just prior to each reading. Remember to use cuvette 1 occasionally to check and adjust the spectrophotometer to 100% transmittance.

10. Obtain the *boiled* chloroplast suspension, mix, and transfer 3 drops to cuvette 4. *Immediately* cover and mix cuvette 4. Insert into the sample holder, read the % transmittance, and record it in Table 4.4. Place cuvette 4 in the incubation test tube rack next to cuvettes 2 and 3. Take and record additional readings at 5, 10, and 15 minutes. Mix the cuvette's contents just prior to each reading. Remember to use cuvette 1 occasionally to check and adjust the spectrophotometer to 100% transmittance.

11. Cover and mix the contents of cuvette 5. Insert it into the sample holder, read the % transmittance, and record it in Table 4.4. Place cuvette 5 in the incubation test tube rack next to tubes 2, 3, and 4. Take additional readings at 5, 10, and 15 minutes. Mix the cuvette's contents just prior to each reading. Remember to use cuvette 1 occasionally to check and adjust the spectrophotometer to 100% transmittance.

Table 4.4: Transmittance (%)

	Time (minutes)			
Cuvette	**0**	**5**	**10**	**15**
2 Unboiled/Dark				
3 Unboiled/Light				
4 Boiled/Light				
5 No Chloroplasts/Light				

Analysis of Results

Plot the percentage of transmittance from the four cuvettes on Graph 4.1. Label each plotted line.

For this graph you will need to determine the following:

a. The *independent* variable: ______________________________
Use this to label the horizontal (x) axis.

b. The *dependent* variable: ______________________________
Use this to label the vertical (y) axis.

Graph 4.1 **Title:** ______________________________

Topics for Discussion

1. What is the function of DPIP in this experiment?

2. What molecule found in chloroplasts does DPIP "replace" in this experiment?

3. What is the source of the electrons that will reduce DPIP?

4. What was measured with the spectrophotometer in this experiment?

5. What is the effect of darkness on the reduction of DPIP? Explain.

6. What is the effect of boiling the chloroplasts on the subsequent reduction of DPIP? Explain.

7. What reasons can you give for the difference in the percentage of transmittance between the live chloroplasts that were incubated in the light and those that were kept in the dark?

8. Identify the function of each of the cuvettes.

Cuvette 1:

Cuvette 2:

Cuvette 3:

Cuvette 4:

Cuvette 5:

LAB FIVE CELL RESPIRATION

OVERVIEW

In this experiment you will work with seeds that are living but **dormant.** A seed contains an embryo plant and a food supply surrounded by a seed coat. When the necessary conditions are met, germination occurs and the rate of cellular respiration greatly increases. In this lab you will:

1. measure oxygen consumption during germination,
2. measure the change in gas volume in respirometers containing either germinating or nongerminating pea seeds, and
3. measure the rate of respiration of these peas at two different temperatures.

OBJECTIVES

Before doing this lab you should understand:

- respiration, dormancy, and germination;
- how a respirometer works in terms of the gas laws;
- the general processes of metabolism in living organisms; and
- how the rate of cellular respiration relates to the amount of activity in a cell.

After doing this lab you should be able to:

- calculate the rate of cell respiration from experimental data;
- relate gas production to respiration rate;
- test the rate of cellular respiration in germinating versus nongerminated seeds in a controlled experiment; and
- test the effect of temperature on the rate of cell respiration in germinating versus nongerminated seeds in a controlled experiment.

INTRODUCTION

Aerobic cellular respiration is the release of energy from organic compounds by metabolic chemical oxidation in the mitochondria within each cell. Cellular respiration involves a series of enzyme-mediated reactions.

The equation below shows the complete oxidation of glucose. Oxygen is required for this energy-releasing process to occur.

$$C_6H_{12}O_6 + 6\ O_2 \rightarrow 6\ CO_2 + 6\ H_2O + 686 \text{ kilocalories of energy/mole of glucose oxidized}$$

By studying the equation above, you will notice there are three ways cellular respiration could be measured. One could measure the

1. **Consumption of O_2** (How many moles of O_2 are consumed in cellular respiration?)
2. **Production of CO_2** (How many moles of CO_2 are produced in cellular respiration?)
3. **Release of energy during cellular respiration**

In this experiment the relative volume of O_2 consumed by germinating and nongerminating (dry) peas at two different temperatures will be measured.

Background Information

A number of physical laws relating to gases are important to the understanding of how the apparatus that you will use in this exercise works. The laws are summarized in the general gas law that states:

$$PV = nRT$$

where **P** is the **pressure** of the gas,

V is the **volume** of the gas,

n is the **number** of molecules of gas,

R is the **gas constant** (its value is fixed), and

T is the **temperature** of the gas (in °K).

This law implies the following important concepts about gases:

1. If the temperature and pressure are kept constant, then the volume of the gas is directly proportional to the number of molecules of the gas.

2. If the temperature and volume remain constant, then the pressure of the gas changes in direct proportion to the number of molecules of gas present.

3. If the number of gas molecules and the temperature remain constant, then the pressure is inversely proportional to the volume.

4. If the temperature changes and the number of gas molecules is kept constant, then either the pressure or volume (or both) will change in direct proportion to the temperature.

It is also important to remember that gases and fluids flow from regions of high pressure to regions of low pressure.

In this experiment the CO_2 produced during cellular respiration will be removed by potassium hydroxide (KOH) and will form solid potassium carbonate (K_2CO_3) according to the following reaction:

$$CO_2 + 2\,KOH \rightarrow K_2CO_3 + H_2O$$

Since the CO_2 is being removed, the change in the volume of gas in the respirometer will be directly related to the amount of oxygen consumed.

In the experimental apparatus shown in Figures 5.1 and 5.2, if water temperature and volume remain constant, the water will move toward the region of lower pressure. During respiration, oxygen will be consumed. Its volume will be reduced, because the CO_2 produced is being converted to a solid. The net result is a decrease in gas volume within the tube and a related decrease in pressure in the tube. The vial with glass beads alone will permit detection of any changes in volume due to atmospheric pressure changes or temperature changes.

The amount of O_2 consumed will be measured over a period of time. Six respirometers should be set up as follows:

Respirometer	Temperature	Contents
1	Room	Germinating Seeds
2	Room	Dry Seeds + Beads
3	Room	Beads
4	10°C	Germinating Seeds
5	10°C	Dry Seeds + Beads
6	10°C	Beads

Procedure

1. Both a room-temperature bath (by convention, 25°C) and a 10°C bath should be set up immediately to allow time for the temperature of each to adjust. Add ice to attain 10°C.

2. **Respirometer 1:** Obtain a 100-mL graduated cylinder and fill it with 50 mL of H_2O. Drop 25 germinating peas in the gradulated cylinder and determine the amount of water that was displaced (which is equivalent to the volume of the peas). Record the volume of the 25 germinating peas. Remove these peas and place them on a paper towel. They will be used in respirometer 1.

   Pea Volume ______________ mL

3. **Respirometer 2:** Refill the graduated cylinder with 50 mL of H_2O. Drop 25 dried peas (not germinating) into the graduated cylinder and then add enough glass beads to attain a volume equivalent to that of the expanded germinating peas. Remove these peas and beads and place them on a paper towel. They will be used in respirometer 2.

4. **Respirometer 3:** Refill the graduated cylinder with 50 mL of H_2O. Determine how many glass beads would be required to attain a volume equivalent to that of the germinating peas. Remove these beads and place them on a paper towel. They will be used in respirometer 3.

5. Repeat Steps 1–4 to prepare a second set of germinating peas, dry peas plus beads, and beads for use in respirometers 4, 5, and 6, respectively.

6. To assemble the six respirometers, obtain six vials, each with an attached stopper and pipette. Place a small piece of cotton in the bottom of each vial and, using a dropper, moisten the cotton with 15% KOH.* Make sure that the respirometer vials are dry on the inside. Do not get KOH on the sides of the respirometer. Place a small wad of nonabsorbent cotton on top of the KOH-soaked absorbent cotton (see Figure 5.1). *It is important that the amounts of cotton and KOH be the same for each respirometer.*

* Your teacher may ask you to use soda-lime pellets instead of KOH Solution.

Figure 5.1: Assembled Respirometers

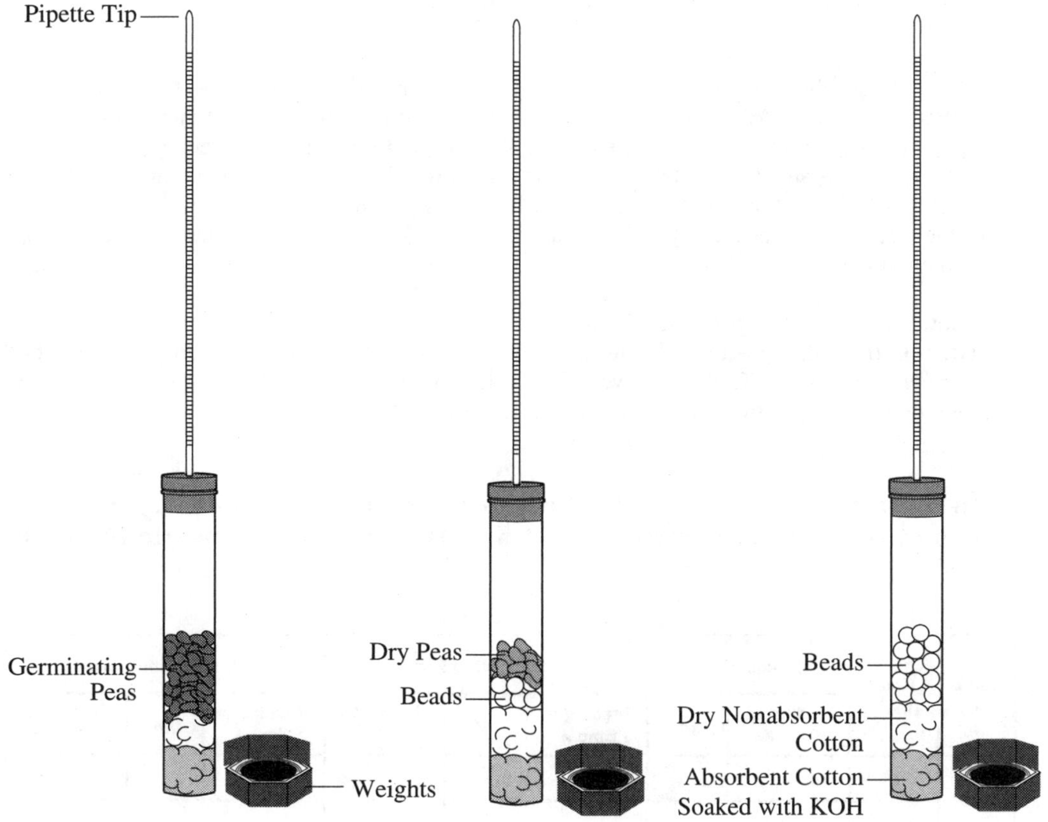

7. Place the first set of germinating peas, dry peas plus beads, and beads in vials 1, 2, and 3, respectively. Place the second set of germinating peas, dry peas plus beads, and beads in vials 4, 5, and 6, respectively. Insert the stopper fitted with the calibrated pipette. Place a weighted collar on each end of the vial (see Figure 5.2).

Figure 5.2: Respirometers Equilibrating in the Water Bath

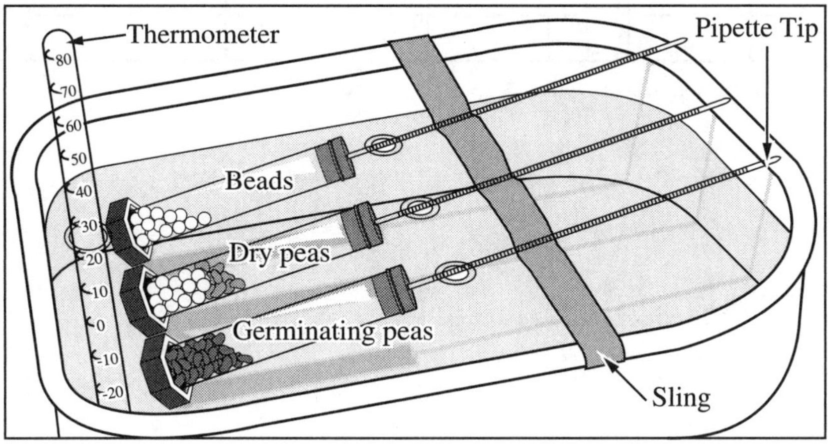

8. Make a sling of masking tape attached to each side of each of the water baths to hold the pipettes out of the water during an equilibration period of seven minutes. Vials 1, 2, and 3 should rest in the room-temperature water bath (approximately 25°C) and vials 4, 5, and 6 should rest in the 10°C water bath (see Figure 5.2).

9. After the equilibration period of seven minutes, immerse all six respirometers *entirely* in their water baths. Water will enter the pipettes for a short distance and then stop. If the water continues to move into a pipette, check for leaks in the respirometer. Work swiftly and arrange the pipettes so that they can be read through the water at the beginning of the experiment. They should not be shifted during the experiment. Hands should be kept out of the water bath after the experiment has started. Make sure that a constant temperature is maintained.

10. Allow the respirometers to equilibrate for three more minutes and then record, to the nearest 0.01 mL, the initial position of water in each pipette (time 0). Check the temperature in both baths and record it in Table 5.1. Every 5 minutes for 20 minutes, take readings of the water's position in each pipette and record the data in Table 5.1.

Table 5.1: Measurement of O_2 Consumption by Soaked and Dry Pea Seeds at Room Temperature (25°C) and 10°C Using Volumetric Methods

Temp (°C)	Time (min)	Beads Alone		Germinating Peas			Dry Peas and Beads		
		Reading at time *X*	Diff.*	Reading at time *X*	Diff.*	Corrected diff.Δ	Reading at time *X*	Diff.*	Corrected diff. Δ
	0								
	5								
	10								
	15								
	20								
	0								
	5								
	10								
	15								
	20								

* Difference = (initial reading at time 0) – (reading at time *X*)
Δ Corrected difference = (initial pea seed reading at time 0 – pea seed reading at time *X*) – (initial bead reading at time 0 – bead reading at time *X*)

Analysis of Results

1. In this activity you are investigating both the effect of germination versus nongermination and warm temperature versus cold temperature on respiration rate. Identify two hypotheses being tested in this activity.

2. This activity uses a number of controls. What conditions must remain constant? Why?

3. Graph the results from the corrected difference column for the germinating peas and the dry peas at both room temperature and at 10°C.

For this graph you will need to determine the following:

a. The *independent* variable: ______________________________
Use this to label the horizontal (x) axis.

b. The *dependent* variable: ______________________________
Use this to label the vertical (y) axis.

Graph 5.1 **Title:** ______________________________

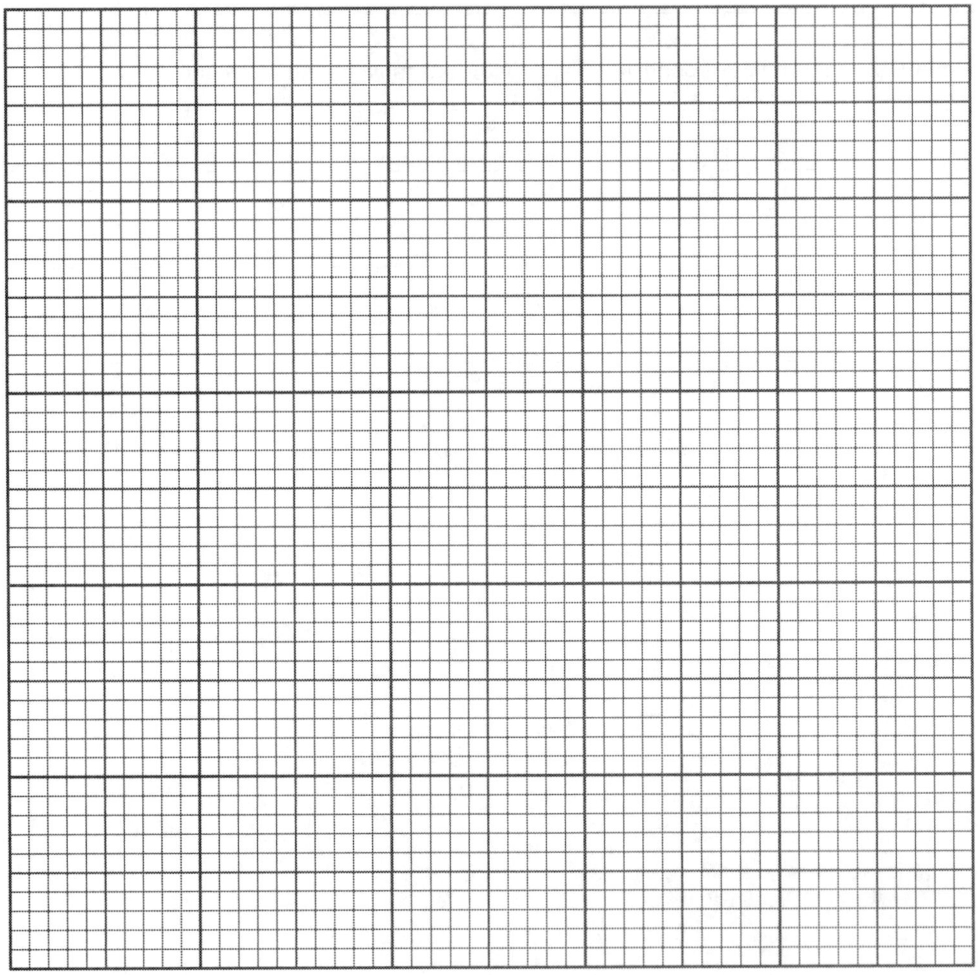

4. Describe and explain the relationship between the amount of O_2 consumed and time.

5. From the slope of the four lines on the graph, determine the rate of O_2 consumption of germinating and dry peas during the experiments at room temperature and at 10°C. Recall that rate = $\frac{\Delta y}{\Delta x}$. Record the rates in Table 5.2.

Table 5.2

Condition	Show Calculations Here	Rate (mL O_2/minute)
Germinating Peas/10°C		
Germinating Peas/ Room Temperature		
Dry Peas/10°C		
Dry Peas/Room Temperature		

6. Why is it necessary to correct the readings from the peas with the readings from the beads?

7. Explain the effect of germination (versus nongermination) on pea seed respiration.

8. Graph 5.2 is a sample graph of possible data obtained for oxygen consumption by germinating peas up to about 8°C. Draw in predicted results through 45°C. Explain your prediction.

Graph 5.2 Title: ______________________________

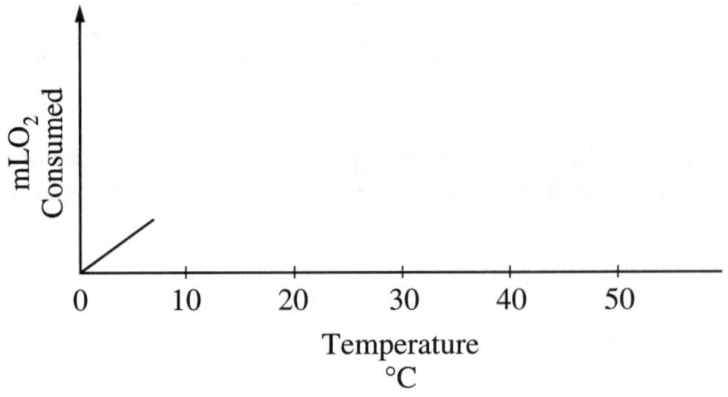

____

____

____

____

9. What is the purpose of KOH in this experiment?

____

____

____

____

____

10. Why did the vial have to be completely sealed around the stopper?

____

____

____

____

____

11. If you used the same experimental design to compare the rates of respiration of a 25 g reptile and a 25 g mammal at 10°C, what results would you expect? Explain your reasoning.

12. If respiration in a small mammal were studied at both room temperature (21°C) and 10°C, what results would you predict? Explain your reasoning.

13. Explain why water moved into the respirometers' pipettes.

14. Design an experiment to examine the rates of cellular respiration with peas that have been germinating for different lengths of time: 0, 24, 48, and 72 hours. What results would you expect? Why?

LAB SIX MOLECULAR BIOLOGY

OVERVIEW

In this lab you will investigate some basic principles of molecular biology:

1. Plasmids containing specific fragments of foreign DNA will be used to transform *Escherichia coli* cells, conferring antibiotic (ampicillin) resistance.
2. Restriction enzyme digests of phage *lambda* DNA will be used to demonstrate techniques for separating and identifying DNA fragments using gel electrophoresis.

OBJECTIVES

Before doing this lab you should understand:

- how gel electrophoresis separates DNA molecules present in a mixture;
- the principles of bacterial transformation;
- the conditions under which cells can be transformed;
- the process of competent cell preparation;
- how a plasmid can be engineered to include a piece of foreign DNA;
- how plasmid vectors are used to transfer genes;
- how antibiotic resistance is transferred between cells;
- how restriction endonucleases function; and
- the importance of restriction enzymes to genetic engineering experiments.

After doing this lab you should be able to:

- use plasmids as vectors to transform bacteria with a gene for antibiotic resistance in a controlled experiment;
- demonstrate how restriction enzymes are used in genetic engineering;
- use electrophoresis to separate DNA fragments;
- describe the biological process of transformation in bacteria;
- calculate transformation efficiency;
- be able to use multiple experimental controls;
- design a procedure to select positively for antibiotic-resistant transformed cells; and
- determine unknown DNA fragment sizes when given DNA fragments of known size.

INTRODUCTION

The bacterium *Escherichia coli* (*E. coli*) is an ideal organism for the molecular geneticist to manipulate and has been used extensively in recombinant DNA research. It is a common inhabitant of the human colon and can easily be grown in suspension culture in a nutrient medium such as Luria broth, or in a petri dish of Luria broth mixed with agar (LB agar) or nutrient agar.

The single circular chromosome of *E. coli* contains about five million DNA base pairs, only 1/600th the haploid amount of DNA in a human cell. In addition, the *E. coli* cell may contain small, circular DNA molecules (1,000 to 200,000 base pairs) called **plasmids,** which also carry genetic information. The plasmids are extrachromosomal; they exist separately from the chromosome. Some plasmids replicate only when the bacterial chromosome replicates and usually exist only as single copies within the bacterial cell. Others replicate autonomously and often occur in as many as 10 to 200 copies within a single bacterial cell. Certain plasmids, called R plasmids, carry genes for resistance to such antibiotics as ampicillin, kanamycin, or tetracycline.

In nature genes can be transferred between bacteria in three ways: conjugation, transduction, or transformation. **Conjugation** is a mating process during which genetic material is transferred from one bacterium to another of a different mating type. **Transduction** requires the presence of a virus to act as a vector (carrier) to transfer small pieces of DNA from one bacterium to another. **Bacterial transformation** involves transfer of genetic information into a cell by direct uptake of the DNA. During gene transfer, the uptake and expression of foreign DNA by a recipient bacterium can result in conferring a particular trait to a recipient lacking that trait. Transformation can occur naturally but the incidence is extremely low and is limited to relatively few bacterial strains. These bacteria can take up DNA only during the period at the end of logarithmic growth. At this time the cells are said to be **competent.** Competence can be induced in *E. coli* with carefully controlled chemical growth conditions. Once competent, the cells are ready to accept DNA that is introduced from another source.

Plasmids can transfer genes (such as those for antibiotic resistance) that occur naturally within them, or plasmids can act as carriers (vectors) for introducing foreign DNA from other bacteria, plasmids, or even eukaryotes into recipient bacterial cells. Restriction endonucleases can be used to cut and insert pieces of foreign DNA into the plasmid vectors (Figure 6.1). If these plasmid vectors also carry genes for antibiotic resistance, transformed cells containing plasmids that carry the foreign DNA of interest in addition to the antibiotic resistance gene can be easily selected from other cells that do not carry the gene for antibiotic resistance.

Figure 6.1: Bacterial Transformation Using a Restriction Endonuclease

I. Create plasmid with gene of interest.

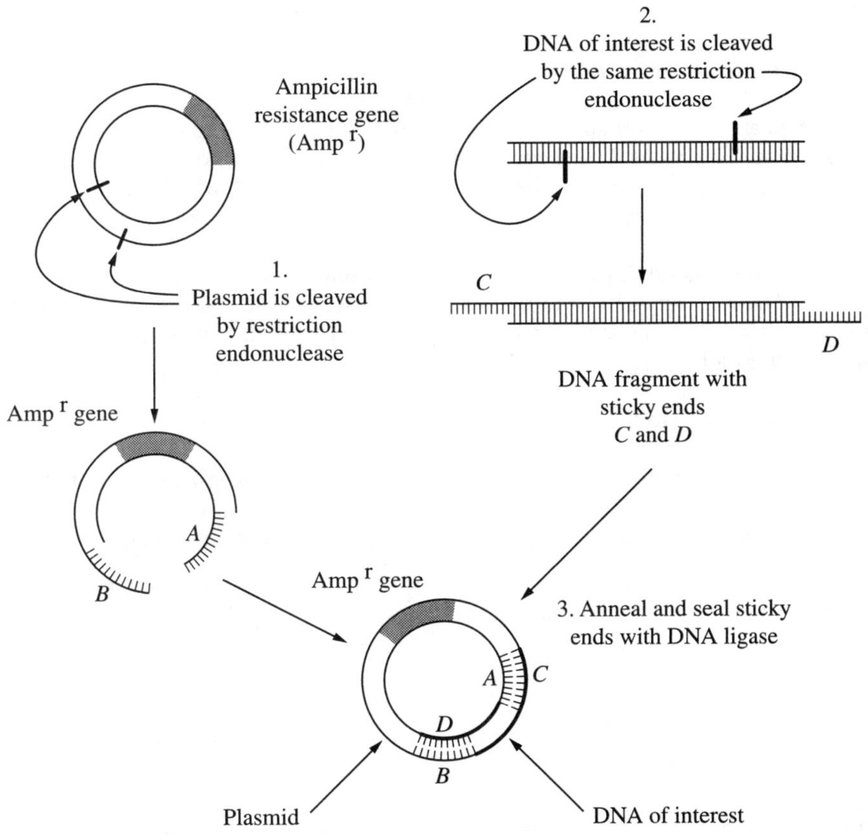

II. Transform recipient cells with plasmid DNA.

III. Plate recipients on ampicillin plates and select for resistant colonies.

IV. Isolate colonies carrying the plasmid.

EXERCISE 6A: Bacterial Transformation—Ampicillin Resistance*

You will insert a plasmid that contains a gene for resistance to ampicillin, an antibiotic that is lethal to many bacteria, into competent *E. coli* cells.

Transformed bacteria can be selected based on their resistance to ampicillin by spreading the transformed cells on nutrient medium that contains ampicillin. Any cells that grow on this medium have been transformed.

Procedure

1. Mark 1 sterile 15-mL tube "+"; this tube will have the plasmid added to it. Mark another tube "–"; this tube will have no plasmid added.

2. Use a sterile micropipette to add 250 microliters (μL) of *ice cold* 0.05 *M* $CaCl_2$ to each tube.

3. Transfer a large (3 mm) colony of *E. coli* from a starter plate to *each* of the tubes using a sterile inoculating loop. Try to get the same amount of bacteria into each tube. Be careful not to transfer any agar.

4. Vigorously tap the loop against the wall of the tube to dislodge the cell mass.

5. Mix the suspension by repeatedly drawing in and emptying a sterile micropipette with the suspension.

6. Add 10 μL of pAMP solution (0.005 μg/μL) directly into the cell suspension in tube "+". Mix by tapping the tube with your finger. This solution contains the antibiotic-resistant plasmid.

7. Keep both tubes on ice for 15 minutes.

8. While the tubes are on ice, obtain 2 LB agar plates and 2 LB/Amp agar (LB agar containing ampicillin) plates. Label each plate on the bottom as follows: one LB agar plate "LB+" and the other "LB–"; label one LB/Amp plate "LB/Amp+" and the other "LB/Amp–".

9. A brief pulse of heat facilitates entry of foreign DNA into the *E. coli* cells. Heat-shock cells in both the "+" and "-" tubes by holding the tubes in a 42°C water bath for 90 seconds. It is essential that cells be given a sharp and distinct shock, so take the tubes directly from the ice to the 42°C water bath.

10. Immediately return cells to the ice for 2 minutes.

11. Use a sterile micropipette to add 250 μL of room-temperature Luria-Bertani broth to each tube. Mix by tapping with your finger. Any transformed cells are now resistant to ampicillin because they possess the gene whose product renders the antibiotic ineffective.

12. Place 100 mL of "+" cells on the "LB+" plate and on the "LB/Amp+" plate. Place 100 mL of "–" cells on the "LB–" plate and on the "LB/Amp–" plate.

---

* Exercise 6A is adapted with permission from *DNA Science: A First Course in Recombinant-DNA Technology*, by David A. Micklos, DNA Learning Center of Cold Spring Harbor Laboratory, and Greg A. Freyer, Columbia University College of Physicians and Surgeons, Copyright 1990 Cold Spring Harbor Laboratory Press and Carolina Biological Supply Company. It is based on a protocol published by Douglas Hanahan, University of California, San Francisco.

13. Immediately spread the cells by using a sterile spreading rod. Repeat the procedure for each plate.

14. Allow plates to set for several minutes. Tape your plates together and incubate *inverted* overnight at 37°C.

Analysis of Results

1. Observe the colonies through the bottom of the culture plate. *Do not open the plates.* Count the number of individual colonies; use a permanent marker to mark each colony as it is counted. If cell growth is too dense to count individual colonies, record "lawn."

   LB + (Positive Control) ________________ LB – (Positive Control) ________________

   LB/Amp + (Experimental) ______________ LB/Amp – (Negative Control) _____________

2. Compare and contrast the number of colonies on each of the following pairs of plates. What does each pair of results tell you about the experiment?

   a. LB + and LB – ____

   ____

   b. LB/Amp – and LB/Amp + ______________________________________

   ____

   c. LB/Amp + and LB + ____

   ____

3. Transformation efficiency is expressed as the number of antibiotic-resistant colonies per microgram of pAMP. Because transformation is limited to only those cells that are competent, increasing the amount of plasmid used does not necessarily increase the probability that a cell will be transformed. A sample of competent cells is usually saturated with small amounts of plasmid, and excess DNA may actually interfere with the transformation process.

   a. Determine the total mass of pAMP used. ______________________________
   (You used 10 μL of pAMP at a concentration of 0.005 μg/μL.)
   Total Mass = volume $\times$ concentration.

   b. Calculate the total volume of cell suspension prepared. ____________________

   c. Now calculate the fraction of the total cell suspension that was spread on the plate.
   Number of μL spread/total volume. ________________________________

   d. Determine the mass of pAMP in cell suspension. ________________________
   Total mass of pAMP X fraction spread.

   e. Determine the number of colonies per mg of plasmid. Express in scientific notation.
   Number of colonies observed/mass pAMP spread [from calculation in Step 3.d] =
   transformation efficiency. ____

4. This is the **transformation efficiency.** What factors might influence transformation efficiency? Explain the effect of each you mention.

____

____

____

____

____

____

EXERCISE 6B: Restriction Enzyme Cleavage of DNA and Electrophoresis

Restriction enzymes, or **restriction endonucleases,** are essential tools in recombinant DNA methodology. Several hundred have been isolated from a variety of prokaryotic organisms. Restriction endonucleases are named according to a specific system of nomenclature. The letters refer to the organism from which the enzyme was isolated. The first letter of the name stands for the genus name of the organism. The next two letters represent the second word, or the species name. The fourth letter (if there is one) represents the strain of the organism. Roman numerals indicate whether the particular enzyme was the first isolated, the second, or so on.

Examples:

HaeIII H = *Haemophilus*

ae = *aegyptus*

III = second endonuclease isolated

EcoRI E = genus *Escherichia*

co = species *coli*

R = strain RY13

I = first endonuclease isolated

Restriction endonucleases recognize specific DNA sequences in double-stranded DNA (usually a four to six base pair sequence of nucleotides) and digest the DNA at these sites. The result is the production of fragments of DNA of various lengths. Some restriction enzymes cut cleanly through the DNA helix at the same position on both strands to produce fragments with blunt ends (Figure 6.2a). Other endonucleases cleave each strand off-center at specific nucleotides to produce fragments with "overhangs," or sticky ends (Figure 6.2b). By using the same restriction enzyme to "cut" DNA from two different organisms, complementary "overhangs," or sticky ends, will be produced and can allow the DNA from two sources to be "recombined." Digestion eith EcoRI or HindIII will produce DNA fragments with sticky ends (Figure 6.2b).

Figure 6.2a

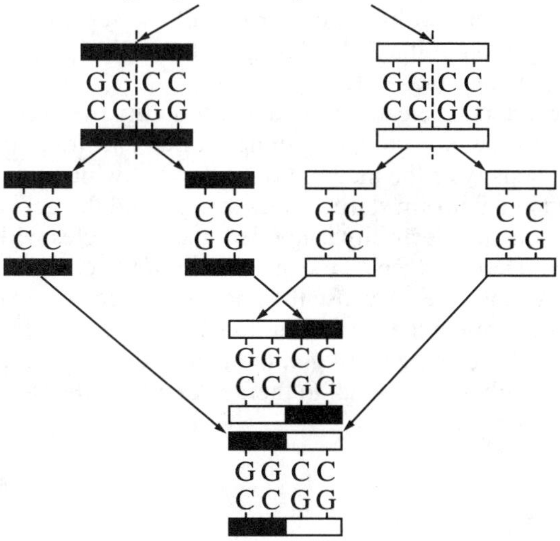

Figure 6.2b

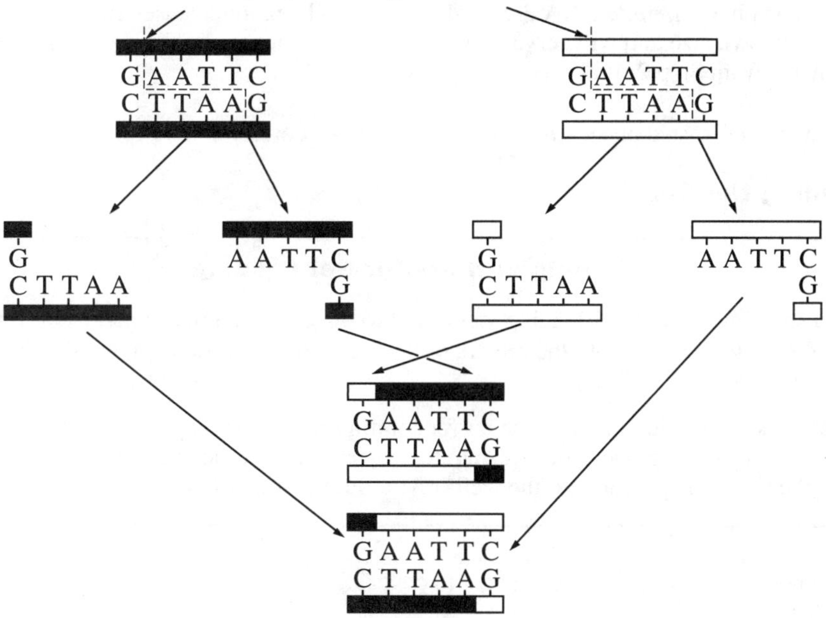

In this exercise samples of DNA obtained from the bacteriophage *lambda* have been incubated with different restriction enzymes. The resulting fragments of DNA will be separated by using gel electrophoresis. One sample has been digested with the restriction endonuclease **EcoRI**, one with the restriction endonuclease **HindIII**, and the third sample is uncut. The DNA samples will be loaded into wells of an agarose gel and separated by the process of electrophoresis. After migration of the DNA through an electrical field, the gel will be stained with methylene blue, a dye that binds to DNA.

When any molecule enters an electrical field, the mobility or speed at which it will move is influenced by the charge of the molecule, the strength of the electrical field, the size and shape of the molecule, and the density of the medium (gel) through which it is migrating. When all molecules are positioned at a uniform starting site on a gel and the gel is placed in a chamber containing a buffer solution and electricity is applied, the molecules will migrate and appear as bands. Nucleic acids, like DNA and RNA, move because of the charged phosphate groups in the backbone of the DNA molecule. Because the phosphates are negatively charged at neutral pH, the DNA will migrate through the gel toward the positive electrode.

In this exercise we will use an **agarose gel.** In agarose the migration rate of linear fragments of DNA is inversely proportional to their size; the smaller the DNA molecule, the faster it migrates through the gel.

Procedure

A: Preparing the Gel

1. Prepare the agarose gel for electrophoresis according to the directions given by your teacher or in the kit.

2. Obtain the phage *lambda* DNA digested with **EcoRI** endonuclease. The DNA is mixed with a gel-loading solution containing a tracking dye, bromophenol blue, that will make it possible to "track" the progress of its migration in the agarose gel.

3. Obtain the phage *lambda* DNA digested with **HindIII** endonuclease. The DNA fragments are of a known size and will serve as a "standard" for measuring the size of the **EcoRI** fragments from Step 2. It also contains the tracking dye.

4. Obtain the undigested phage *lambda* DNA to use as a control. It also contains the tracking dye.

B: Loading the Gel

Helpful Hints for Gel Loading

Pull a small amount of gel-loading solution into the end of a micropipette. (Do not allow the solution to move up into the pipette or bubbles will be introduced into the well of the agarose gel during loading.)

Hold the tip of the pipette in the buffer solution *above* the well and gently dispense the solution. The loading dye is denser than the buffer and will move into the well. (Do *not* place the tip of the pipette into the well or you might puncture the gel).

1. Pour enough buffer gently over the gel to cover it.

2. Load 5–10 μL of undigested *lambda* phage DNA (control) into a well.

3. Load 5–10 μL of the **HindIII** digest into a second well.

4. Load 5–10 μL of the **EcoRI** digest into a third well.

C: Electrophoresis

1. Place the top on the electrophoresis chamber and carefully connect the electrical leads to an approved power supply (black to black and red to red). Set the voltage to the appropriate level for your apparatus. When the current is flowing, you should see bubbles on the electrodes.

2. Allow electrophoresis to proceed until the tracking dye has moved nearly to the end of the gel.

3. After electrophoresis is completed, *turn off the power*, disconnect the leads, and remove the cover of the electrophoresis chamber.

D: Staining and Visualization

Note: Wear gloves.

1. Carefully remove the gel bed from the chamber and gently transfer the gel to a staining tray for staining. Use the scooper provided with your kit or keep your hands under the gel during the transfer. You may wish to remove a small piece of gel from the upper right-hand corner to keep track of the gel's orientation. *Do not stain in the electrophoresis chamber.*

2. Label the staining tray with your name and take it to your teacher for staining.

3. Examine your stained gel on a light box or overhead projector. Compare your gel with the sample gel shown in Figure 6.3.

Figure 6.3: Sample Restriction Digest of *Lambda* DNA

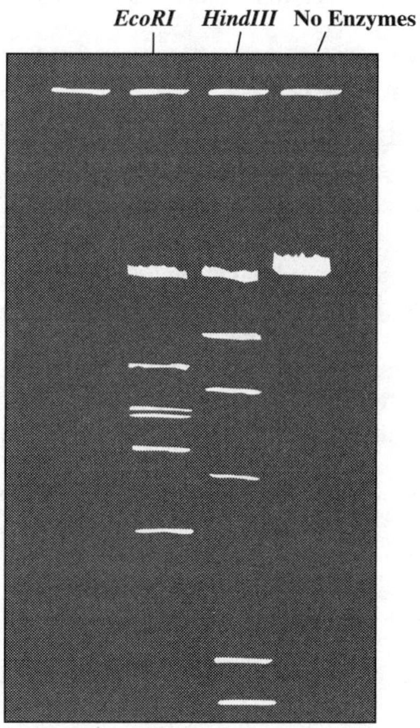

Figure not to scale.

E: Determining Fragment Size

1. After observing the gel, carefully wrap it in plastic wrap and smooth out all of the wrinkles.

2. Using a marking pen, trace the outlines of the sample wells and the location of the bands.

3. Remove the plastic wrap and flatten it out on a white piece of paper on the lab bench. Save the gel in a Ziploc® plastic bag. Add several drops of buffer. Store at 4°C. You can make your measurements directly from the marked plastic wrap.

Analysis and Results

The size of the fragments produced by a specific endonuclease (**EcoRI** in this exercise) can be determined by using standard fragments of known size (fragments produced by **HindIII,** in this case). When you plot the data on semilog graph paper, the size of the fragments is expressed as the log of the number of base pairs they contain. This allows data to be plotted on a straight line. The migration distance of the unknown fragments, plotted on the x-axis, will allow their size to be determined on the standard curve.

Graphing

A. Standard Curve for HindIII

1. Measure the migration distance (in cm) for each **HindIII** band on your gel. Measure from the bottom of the sample well to the bottom of the band. The migration distance for the largest standard fragment (approximately 23,120 base pairs) nearest to the origin does not need to be measured. Record these measurements in Table 6.1.

2. Plot the measured migration distance for each band of the standard **HindIII** digest against the actual base pair (bp) fragment sizes given in Table 6.1 using the semilog graph paper of Graph 6.1. Follow your teacher's directions to draw the best-fit line to your points. This will serve as a **standard curve.**

B. Interpolated Calculations for EcoRI

From the standard curve for **HindIII,** made from known fragment sizes, you can calculate fragment sizes resulting from a digest with **EcoRI.** The procedure is as follows.

1. Measure the migration distances in cm for each **EcoRI** band. Record the data in Table 6.2.

2. Determine the sizes of the fragments of phage *lambda* DNA digested with **EcoRI.** Locate on the x-axis of Graph 6.1 the distance migrated by the first **EcoRI** fragment. Using a ruler, draw a vertical line from this point to its intersection with the best-fit data line. Now extend a horizontal line from the intersection point to the y-axis. This point gives the base pair size of this **EcoRI** fragment. Repeat this procedure and determine the remaining **EcoRI** fragments. Enter your interpolated data in Table 6.2, in the interpolated bp column.

3. Your teacher will provide you with the actual bp data. Compare your results to these actual sizes.

Note: This interpolation technique is not exact. You should expect as much as 10% to 15% error.

Table 6.1: Distance *HindIII* Produced Fragments Migrate in Agarose Gel (cm)

HindIII	
Actual bp	Measured Distance (cm)
23,130	—
9,416	______
6,557	______
4,361	______
2,322*	______
2,027*	______
570*Δ	______
125 D	______

Table 6.2: Distance *EcoRI* Produced Fragments Migrate in Agarose Gel (cm)

	EcoRI		
	Measured Distance (cm)	Interpolated bp	Actual bp
Band 1	______	______	______
Band 2	______	______	______
Band 3	______	______	______
Band 4	______	______	______
Band 5	______	______	______
Band 6	______	______	______

* may form a single band
Δ may not be detected

Graph 6.1 Title: ______________________________

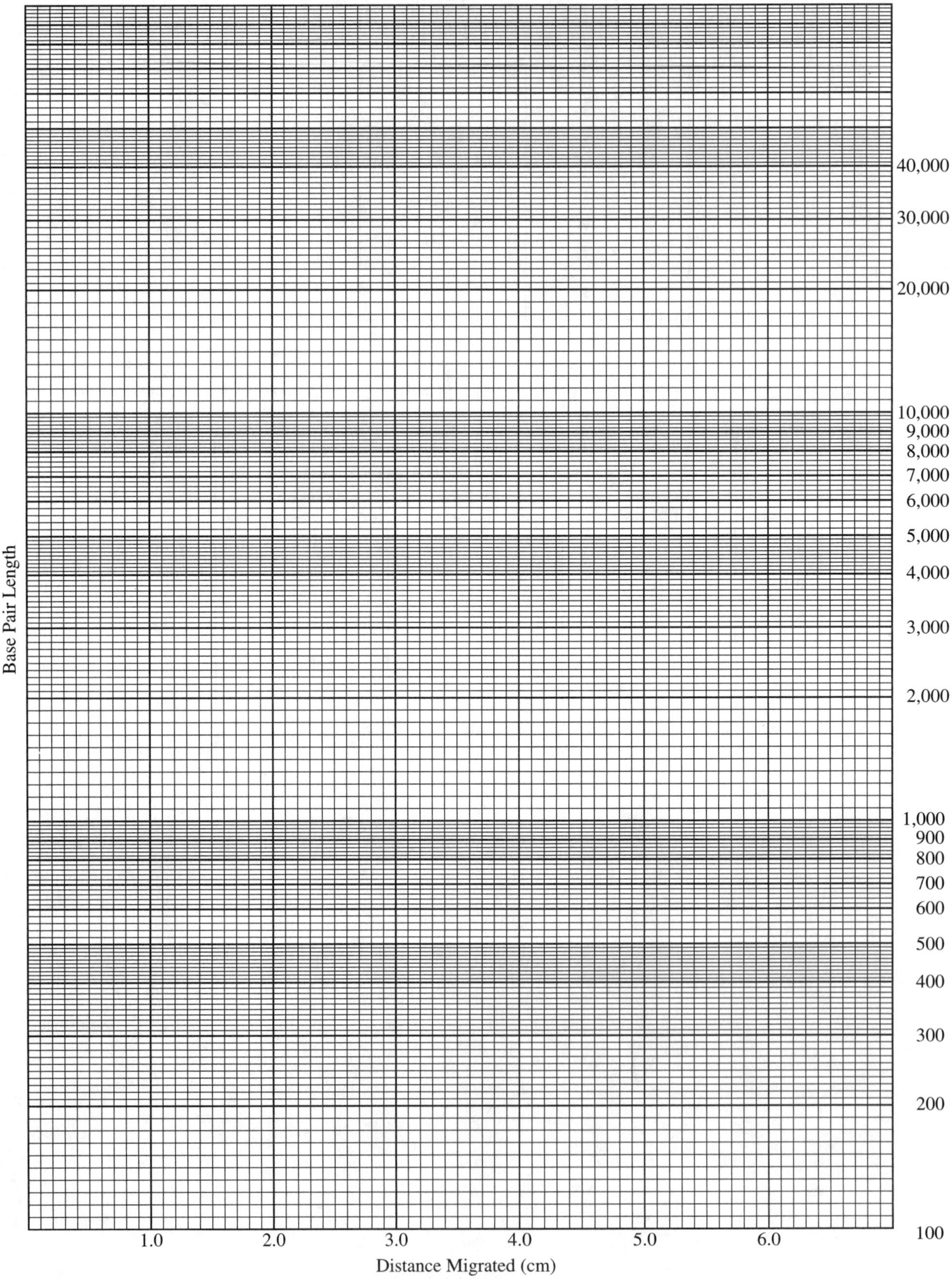

4. For which fragment sizes was your graph most accurate? For which fragment sizes was it least accurate? What does this tell you about the resolving ability of agarose-gel electrophoresis?

Analysis

1. Discuss how each of the following factors would affect the results of electrophoresis:

a. Voltage used

b. Running time

c. Amount of DNA used

d. Reversal of polarity

2. Two small restriction fragments of nearly the same base pair size appear as a single band, even when the sample is run to the very end of the gel. What could be done to resolve the fragments? Why would it work?

Questions

1. What is a plasmid? How are plasmids used in genetic engineering?

2. What are restriction enzymes? How do they work? What are recognition sites?

___

___

___

3. What is the source of restriction enzymes? What is their function in nature?

___

___

___

4. Describe the function of electricity and the agarose gel in electrophoresis.

___

___

___

___

___

5. A certain restriction enzyme digest results in DNA fragments of the following sizes: 4,000 base pairs, 2,500 base pairs, 2,000 base pairs, 400 base pairs. Sketch the resulting separation by electrophoresis. Show starting point, positive and negative electrodes, and the resulting bands.

6. What are the functions of the loading dye in electrophoresis? How can DNA be prepared for visualization?

___

___

___

___

___

7. Use the graph you prepared from your lab data to predict how far (in cm) a fragment of 8,000 bp would migrate.

8. How can a mutation that alters a recognition site be detected by gel electrophoresis?

LAB SEVEN GENETICS OF ORGANISMS

OVERVIEW

In this lab you will use living organisms to do genetic crosses. You will learn how to collect and manipulate the organisms, collect data from F_1 and F_2 generations, and analyze the results from a monohybrid, dihybrid, or sex-linked cross. The procedures that follow apply to fruit flies; your teacher may substitute other procedures using other organisms.

OBJECTIVES

Before doing this lab you should understand:

- chi-square analysis of data, and
- the life cycle of diploid organisms useful in genetics studies.

After doing this lab you should be able to:

- investigate the independent assortment of two genes and determine whether the two genes are autosomal or sex-linked using a multigeneration experiment, and
- analyze the data from your genetic crosses using chi-square analysis techniques.

INTRODUCTION

Drosophila melanogaster, the fruit fly, is an excellent organism for genetics studies because it has simple food requirements, occupies little space, is hardy, completes its life cycle in about 12 days at room temperature, produces large numbers of offspring, can be immobilized readily for examination and sorting, and has many types of hereditary variations that can be observed with low-power magnification. *Drosophila* has a small number of chromosomes (four pairs). These chromosomes are easily located in the large salivary gland cells. *Drosophila* exists in stock cultures that can be readily obtained from several sources. Much research about the genetics of *Drosophila* during the last 50 years has resulted in a wealth of reference literature and a knowledge about hundreds of its genes.

The Life Cycle of *Drosophila*

The Eggs. The eggs are small, oval shaped, and have two filaments at one end. They are usually laid on the surface of the culture medium and, with practice, can be seen with the naked eye. The eggs hatch into larvae after about a day.

The Larval Stage. The wormlike larva eats almost continuously, and its black mouth parts can easily be seen moving back and forth even when the larva itself is less distinct. Larvae tunnel through the culture medium while eating; thus, channels are a good indication of the successful growth of a culture. The larva sheds its skin twice as it increases in size. In the last of the three larval stages, the cells of the salivary glands contain giant chromosomes, which may be seen readily under low-power magnification after proper staining.

The Pupal Stage. When a mature larva in a lab culture is about to become a pupa, it usually climbs up the side of the culture bottle or on to the strip provided in the culture bottle. The last larval covering then becomes harder and darker, forming the pupal case. Through this case the later stages of metamorphosis to an adult fly can be observed. In particular, the eyes, the wings, and the legs become readily visible.

The Adult Stage. When metamorphosis is complete, the adult flies emerge from the pupal case. They are fragile and light in color and their wings are not fully expanded. These flies darken in a few hours and take on the normal appearance of an adult fly. They live a month or more and then die. A female does not mate for about ten to twelve hours after emerging from the pupa. Once she has mated, she stores a considerable quantity of sperm in receptacles and fertilizes her eggs as she lays them. To ensure a controlled mating, it is necessary to use females that have not mated before (virgins).

Figure 7.1: The Life Cycle of *Drosophila melanogaster*

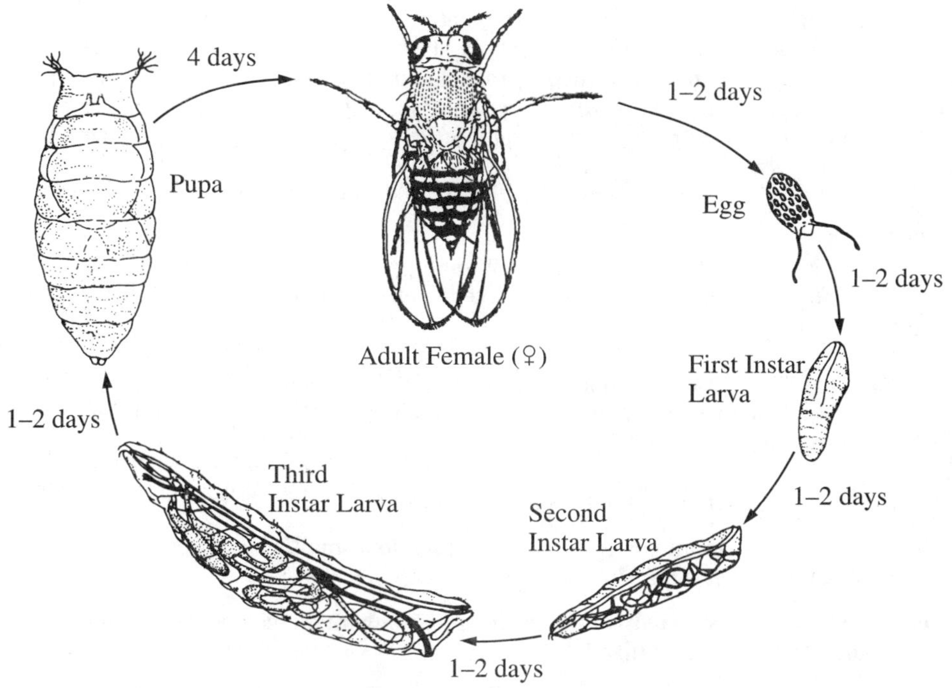

It is important to realize that a number of factors determine the length of time of each stage in the life cycle. Of these factors, temperature is the most important. At room temperature (about 25°C), the complete cycle takes ten to twelve days.

Design of the Exercise

This genetics experiment will be carried on for several weeks. *Drosophila* with well-defined mutant traits will be assigned to you by your teacher. You are responsible for making observations and keeping records concerning what happens as mutant traits are passed from one generation to the next.

You will be assigned to study a certain mode of inheritance using particular genetic crosses of flies having one or two mutations. The modes of inheritance most commonly used are:

1. **Monohybrid.** In these experiments the mode of inheritance is determined when a single contrasting pair of characteristics is involved.

2. **Dihybrid.** In these experiments the mode of inheritance is determined when two pairs of contrasting characteristics are considered simultaneously.

3. **Sex-linked.** In these experiments the mode of inheritance is determined when the mutant characteristic is associated with the X-chromosome.

To make these experiments interesting and challenging, you will not be told the mode of inheritance, nor the name for the particular mutation(s) you are studying. Study the wild type flies (both male and female) until their phenotypic characteristics are familiar. Flies having one or two mutations can then be identified by making comparisons with the wild type flies. The most commonly studied mutations are in eye color or shape, bristle number or shape, wing size or shape, or antenna size or shape. You should make up your own name for the particular mutation(s) that you identify in your flies.

Procedure

1. Obtain a vial of wild type flies. Practice immobilizing and sexing (determining the gender of) these flies. Examine these flies and note the characteristics of their eyes, wings, bristles, and antennae.

2. To make handling easier, immobilize the flies by chilling them. Since the activity level of the flies is dependent on environmental temperature, the following steps immobilize the flies. (Your teacher may assign a different method.)

   a. Hold the vial containing the flies at an angle and twirl it in ice for several minutes.

   b. When the flies are immobilized, dump them into a small, plastic petri dish containing a #1 Whatman filter paper.

   c. Place the petri dish on top of the ice in order to maintain the cool temperature necessary to keep the flies immobilized.

   d. Use the dissecting microscope to view the flies. The top of the petri dish can be on or off when viewing.

3. Distinguish male flies from female flies by looking for the following characteristics (illustrated in Figure 7.2):

   a. Males are usually smaller than females.

   b. Males have dark, blunt abdomens, and females have lighter, pointed abdomens.

   c. Only the males have sex combs, which are groups of black bristles on the uppermost joint of the forelegs.

Figure 7.2: Female and Male *Drosophila*

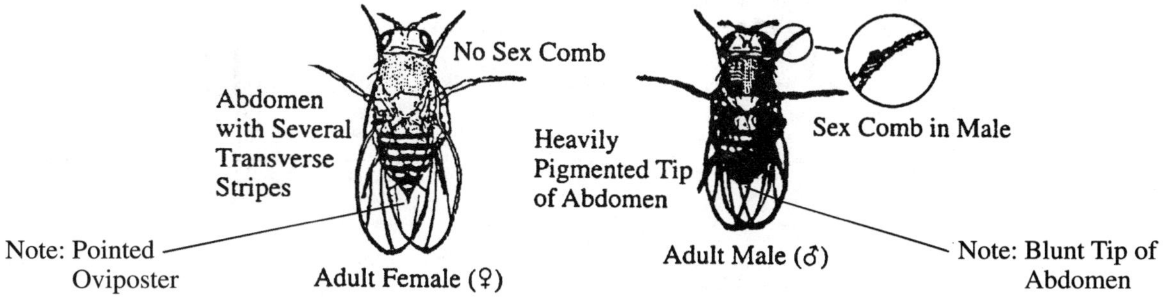

4. Obtain a vial containing pairs of experimental flies. Record the cross number of the vial. This number will serve as a record as to which cross you have obtained. These flies are the parental generation (P) and have already mated. The females should have already laid eggs on the surface of the culture medium. The eggs (or maybe larvae now) represent the first filial, F_1, generation and will be emerging from their pupal cases in about a week.

5. **First Week (Today).** Immobilize and remove the adult flies. Observe them carefully under the dissecting microscope. Separate the males from the females and look for the mutation(s). Note whether the mutation(s) is/are associated with the males or the females. Identify the mutation(s) and give it/them a made-up name and symbol. Record the phenotype and symbol in Table 7.1. The findings should be confirmed by your teacher.

6. Place the parents in the morgue (a jar containing alcohol or baby oil). Label the vial containing the eggs or larvae with symbols for the mating. For example, if a sepia-eyed female is crossed with a wild-type male, the label could be "sepia ♀ X wild ♂". Also be sure to label the vial with your name and the date. Place the vial in a warm location.

7. **Second Week.** Begin by observing the F_1 flies. Immobilize and examine all the flies. Record their sex and the presence or absence of the mutation(s) (as observed in the parental flies) in Table 7.1. Consider the conclusions that can be drawn from these data. Place 5 or 6 pairs of F_1 flies in a fresh culture bottle and the rest of the flies in the morgue. For this cross the females need not be virgins. Label the new vial "F_1 X F_1". Also, label the vial with symbols denoting the cross, the date, and your name.

8. **Third Week.** Remove the F_1 flies from the vials and place them in the morgue. The F_2 generation are the eggs and/or larvae in the vial. Place the vial back in the warm location.

9. **Fourth Week.** Begin removing the F_2 flies. Record their sex and the presence or absence of the mutant phenotypes (as observed in the parental flies in Table 7.2). The more F_2 flies collected, the more reliable the data will be. You may have to collect flies over a 3- or 4-day period. Try to collect at least 200 flies.

10. To analyze your data, you will need to learn how to use the chi-square test. Go to the Statistical Analysis Section (page 85) to review this technique.

Table 7.1: F_1 Generation Data

Date ______________

Phenotype and Symbol	Females	Males

Table 7.2: F_2 Generation Data

Date ______________

Phenotype and Symbol	Females	Males

Analysis of Results

1. Describe and name the observed mutation(s).

____

____

____

____

____

2. Write a hypothesis that describes the mode of inheritance of the trait(s) you studied. This is your null hypothesis (as described in the Statistical Analysis Section.)

____

____

____

3. Refer to a textbook and review Punnett squares. In the space below construct two Punnett squares to predict the expected results of both the parental and F_1 crosses from your null hypothesis.

Parental Cross	F_1 Cross

4. Refer to the Punnet Squares above. In the box below record the *expected* ratios for the genotypes and phenotypes of the F_1 and F_2 crosses in the experiment.

	Expected Genotypic Ratio	Expected Phenotypic Ratio
F_1		
F_2		

5. Do the actual results deviate from what was expected? If so, explain how.

____

____

____

6. From the results, describe your cross.

Is the mutation sex-linked or autosomal? ________________

Is the mutation a dominant or recessive? ________________

Is the cross monohybrid or dihybrid? ________________

7. Are the deviations for the phenotypic ratio of the F_2 generation within the limits expected by chance? To answer this question, statistically analyze the data using the chi-square analysis. Calculate the chi-square statistic for the F_2 generation in the chart below. Refer to the critical values of the chi-square (χ^2) distribution table (Table 7.5 on page 86) to determine the p (probability) value that is associated with your χ^2 statistic.

Phenotype	# Observed (o)	# Expected (e)	(o − e)	$(o-e)^2$	$\frac{(o-e)^2}{e}$
				$\chi^2 =$	

a. Calculate the chi-square value for these data.

1. How many degrees of freedom are there? ____________

2. chi-square (χ^2) = ____________

3. Referring to the critical values chart, what is the probability value for this data?

____

b. According to the probability value, can you accept or reject your null hypothesis? Explain why.

____

____

____

____

____

Topics for Discussion

1. Why was it necessary for the females of the parental generation to be virgins?

____

____

____

2. Why was it not necessary to isolate virgin females for the F_1 cross?

____

____

____

____

3. Why were the adult flies removed from the vials at weeks 2 and 4?

____

____

____

____

LAB SEVEN STATISTICAL ANALYSIS SECTION

Using the Chi-Square Test for Statistical Analysis of Experimental Data

Example 1

Statistics can be used to determine if differences among groups are significant, or simply the result of predictable error. The statistical test most frequently used to determine whether data obtained experimentally provide a good fit or approximation to the expected or theoretical data is the chi-square test. This test can be used to determine if deviations from the expected values are due to chance alone, or to some other circumstance. For example, consider corn seedlings resulting from an F_1 cross between parents that are heterozygous for color.

A Punnett square of the F_1 cross **Gg X Gg** would predict that the expected proportion of green:albino seedlings would be 3:1. Use this information to fill in the Expected (e) column and the (o−e) column in Table 7.3.

Table 7.3

Phenotype	Genotype	# Observed (o)	# Expected (e)	(o−e)
Green	GG or Gg	72		
Albino	gg	12		
	Total:	84		

There is a small difference between the observed and expected results, but are these data close enough that the difference can be explained by random chance or variation in the sample?

To determine if the observed data fall within acceptable limits, a chi-square analysis is performed to test the validity of a **null hypothesis** (that there is no statistically significant difference between the observed and expected data). If the chi-square analysis indicates that the data vary too much from the expected 3:1, an **alternative hypothesis** is accepted.

The formula for chi-square is:

$$\chi_2 = \Sigma \frac{(o-e)^2}{e}$$

where **o** = **observed** number of individuals

e = **expected** number of individuals

Σ = the **sum of the values** (in this case, the differences, squared, divided by the number expected)

1. This statistical test will examine the null hypothesis, which predicts that the data from the experimental cross above will be expected to fit the 3:1 ratio.
2. Use the data from Table 7.3 to complete Table 7.4.

Table 7.4

Phenotype	# Observed (o)	# Expected (e)	(o−e)	$(o-e)^2$	$\frac{(o-e)^2}{e}$
Green	72				
Albino	12				
				$\chi^2 = \Sigma \frac{(o-e)^2}{e} =$	

3. Your calculations should give you a value for $\chi^2 = 5.14$. This value is then compared to Table 7.5.

Table 7.5: Critical Values of the Chi-Square Distribution

	Degrees of Freedom (df)				
Probability (p)	1	2	3	4	5
0.05	3.84	5.99	7.82	9.49	11.1
0.01	6.64	9.21	11.3	13.2	15.1
0.001	10.8	13.8	16.3	18.5	20.5

How to Use the Critical Values Table

1. Determine the **degrees of freedom (df)** for your experiment. It is the number of phenotypic classes minus 1. Since there are two possible genotypes, for this experiment df = 1 (2 samples –1). If the experiment had gathered data for a dihybrid cross, there would be four possible phenotypes and therefore 3 degrees of freedom.

2. Find the p value. Under the 1 df column, find the critical value in the probability (p) = 0.05 row: it is 3.84. What does this mean? **If the calculated chi-square value is greater than or equal to the critical value from the table, then the null hypothesis is rejected.** Since for our example $\chi^2 = 5.14$ and $5.14 > 3.84$, we reject our null hypothesis that there is no statistically significant difference between the observed and expected data. In other words, chance alone cannot explain the deviations we observed and there is, therefore, reason to doubt our original hypothesis (or to question our data collection accuracy.) The minimum probability for rejecting a null hypothesis in the sciences is generally 0.05, so this is the row to use in the chi-square table.

3. These results are said to be significant at a probability of $p = 0.05$. This means that only 5% of the time would you expect to see similar data if the null hypothesis was correct; thus, you are 95% sure that the data do not fit a 3:1 ratio.

4. Since these data do not fit the expected 3:1 ratio, you must consider reasons for this variation. Additional experimentation would be necessary. Perhaps the sample size is too small, or errors were made in data collection. In this example, perhaps the albino seedlings are underrepresented because they died before the counting was performed.

Example 2

In a study of incomplete dominance in tobacco seedlings, the counts in Table 7.6 were made from a cross between two heterozygous (Gg) plants:

Table 7.6

Phenotype	Genotype	# Observed (0)
Green	GG	22
Yellow Green	Gg	50
Albino	gg	12
	Total:	84

A Punnett square for this cross indicates that the expected counts should be in a 1 green:2 yellow green:1 albino ratio (Table 7.7). The expected values for a total count of 84 organisms are therefore:

1 green $= \frac{1}{4} \times 84 = 21$

2 yellow green $= \frac{1}{2} \times 84 = 42$

1 yellow $= \frac{1}{4} \times 84 = \underline{21}$

84

Table 7.7

Phenotype	# Observed (o)	# Expected (e)	(o − e)	$(o-e)^2$	$\frac{(o-e)^2}{e}$
Green	22	21	1	1	0.05
Yellow Green	50	42	8	64	1.52
Albino	12	21	9	81	3.86
				$\chi^2 = \Sigma\frac{(o-e)^2}{e} =$	5.43

Go to the chi-square table, this time for two degrees of freedom (there are three phenotypes: 3 – 1 = 2 df). If the χ^2 value were greater than or equal to the critical value of 5.99 we would reject our hypothesis. Since 5.43 is less than the critical value at $p = .05$, we accept the null hypothesis (this second data set does fit the expected 1:2:1 ratio).

Practice Problem

An investigator observes that when pure-breeding, long-wing *Drosophila* are mated with pure-breeding, short-wing flies, the F_1 offspring have an intermediate wing length.

When several intermediate-wing-length flies are allowed to interbreed the following results are obtained:

Observed

230 long wings
510 intermediate-length wings
260 short wings

a. What is the genotype of the F_1 intermediate-wing-length flies? ____________________

b. Write a hypothesis describing the mode of inheritance of wing length in *Drosophila* (this is your null hypothesis).

____

____

____

c. Complete Table 7.8.

Table 7.8

Phenotype	# Observed (o)	# Expected (e)	(o − e)	$(o-e)^2$	$\frac{(o-e)^2}{e}$
				$\chi^2 = \Sigma\frac{(o-e)^2}{e} =$	

(i) Calculate the chi-square value for these data.

1. How many degrees of freedom (df) are there? ___________
2. χ^2 (chi-square) = ___________
3. Referring to the critical values chart, what is the probability value for these data?

   ____

(ii) According to the critical value of χ^2, can you accept or reject the null hypothesis? Explain why.

____

____

____

____

____

LAB EIGHT POPULATION GENETICS AND EVOLUTION

OVERVIEW

In this lab you will:

1. learn about the Hardy-Weinberg law of genetic equilibrium, and
2. study the relationship between evolution and changes in allele frequency by using your class to represent a sample population.

OBJECTIVES

Before doing this lab you should understand:

- how natural selection can alter allelic frequencies in a population;
- the Hardy-Weinberg equation and its use in determining the frequency of alleles in a population; and
- the effects on allelic frequencies of selection against the homozygous recessive or other genotypes.

After doing this lab you should be able to:

- calculate the frequencies of alleles and genotypes in the gene pool of a population using the Hardy-Weinberg formula, and
- discuss natural selection and other causes of microevolution as deviations from the conditions required to maintain Hardy-Weinberg equilibrium.

INTRODUCTION

In 1908 G. H. Hardy and W. Weinberg independently suggested a scheme whereby evolution could be viewed as changes in the frequency of alleles in a population of organisms. In this scheme, if *A* and *a* are alleles for a particular gene locus and each diploid individual has two such loci, then p can be designated as the frequency of the *A* allele and q as the frequency of the *a* allele. Thus, in a population of 100 individuals (each with two loci) in which 40% of the alleles are *A*, p would be 0.40. The rest of the alleles (60%) would be *a*, and q would equal 0.60 (i.e., $p + q = 1.0$). These are referred to as **allele frequencies.** The frequency of the possible diploid combinations of these alleles (*AA, Aa, aa*) is expressed as $p^2 + 2\,pq + q^2 = 1.0$. Hardy and Weinberg also argued that if five conditions are met, the population's allele and genotype frequencies will remain constant from generation to generation. These conditions are as follows:

1. **The breeding population is large.** (The effect of chance on changes in allele frequencies is thereby greatly reduced.)
2. **Mating is random.** (Individuals show no mating preference for a particular phenotype.)
3. **There is no mutation of the alleles.** (No alteration in the DNA sequence of alleles.)
4. **No differential migration occurs.** (No immigration or emigration.)
5. **There is no selection.** (All genotypes have an equal chance of surviving and reproducing.)

The Hardy-Weinberg equation describes an existing situation. If the five conditions are met, then no change will occur in either allele or genotype frequencies in the population. Of what value is such a rule? It provides a yardstick by which changes in allele frequency, and therefore evolution, can be measured. One can look at a population and ask: Is evolution occurring with respect to a particular gene locus? Since evolution is difficult (if not impossible) to observe in most natural populations, we will model the evolutionary process using the class as a simulated population. The purpose of this simulation is to provide an opportunity to test some of the basic tenets of population genetics and evolutionary biology.

EXERCISE 8A: Estimating Allele Frequencies for a Specific Trait within a Sample Population

Using the class as a sample population, the allele frequency of a gene controlling the ability to taste the chemical PTC (phenylthiocarbamide) could be estimated. A bitter-taste reaction to PTC is evidence of the presence of a *dominant allele* in either the homozygous condition (*AA*) or the heterozygous condition (*Aa*). The inability to taste the chemical at all depends on the presence of homozygous recessive alleles (*aa*). (Instead of PTC tasting, other traits, such as attached earlobes, may be used.) To estimate the frequency of the PTC-tasting allele in the population, one must find p. To find p, one must first determine q (the frequency of the nontasting PTC allele), because only the genotype of the homozygous recessive individuals is known for sure (i.e., those that show the dominant trait could be *AA* or *Aa*).

Procedure

1. Using the PTC taste-test papers provided, tear off a short strip and press it to your tongue tip. PTC tasters will sense a bitter taste. For the purposes of this exercise these individuals are considered to be tasters.

2. A decimal number representing the frequency of tasters ($p^2 + 2pq$) should be calculated by dividing the number of tasters in the class by the total number of students in the class. A decimal number representing the frequency of nontasters (q^2) can be obtained by dividing the number of nontasters by the total number of students. You should then record these numbers in Table 8.1.

3. Use the Hardy-Weinberg equation to determine the frequencies (p and q) of the two alleles. The frequency q can be calculated by taking the square root of q^2. Once q has been determined, p can be determined because $1 - q = p$. Record these values in Table 8.1 for the class and also calculate and record values of p and q for the North American population.

Table 8.1: Phenotypic Proportions of Tasters and Nontasters and Frequencies of the Determining Alleles

	Phenotypes				Allele Frequency Based on the H-W Equation	
	Tasters ($p^2 + 2pq$)		Nontasters (q^2)		p	q
Class Population	#	%	#	%		
North American Population	0.55		0.45			

Topics for Discussion

1. What is the percentage of heterozygous tasters (2 pq) in your class? ____________________

2. What percentage of the North American population is heterozygous for the taster trait?

____

EXERCISE 8B: Case Studies

CASE I — A Test of an Ideal Hardy-Weinberg Population

The entire class will represent a breeding population, so find a large open space for this simulation. In order to ensure random mating, choose another student at random. In this simulation, we will assume that gender and genotype are irrelevant to mate selection.

The class will simulate a population of randomly mating heterozygous individuals with an initial gene frequency of 0.5 for the dominant allele *A* and the recessive allele *a* and genotype frequencies of 0.25 *AA,* 0.50 *Aa,* and 0.25 *aa.* Your initial genotype is *Aa.** Record this on the Data Page (page 98). Each member of the class will receive four cards. Two cards will have *A* written on them and two cards will have *a.* The four cards represent the products of meiosis. Each "parent" contributes a haploid set of chromosomes to the next generation.

Procedure

1. Turn the four cards over so that the letters do not show, shuffle them, and take the card on top to contribute to the production of the first offspring. Your partner should do the same. Put the two cards together. The two cards represent the alleles of the first offspring. One of you should record the genotype of this offspring in the Case I section on the Data Page. Each student pair must produce two offspring, so all four cards must be reshuffled and the process repeated to produce a second offspring.

2. The other partner should then record the genotype of the second offspring on the Data Page. The very short reproductive career of this generation is over. You and your partner now become the next generation by assuming the genotypes of the *two offspring*. That is, Student 1 assumes the genotype of the first offspring and Student 2 assumes the genotype of the second offspring.

3. Each student should obtain, if necessary, new cards representing the alleles in his or her respective gametes after the process of meiosis. For example, Student 1 becomes genotype *Aa* and obtains cards *A*, *A*, *a*, *a*; Student 2 becomes *aa* and obtains cards *a*, *a*, *a*, *a*. Each participant should *randomly* seek out another person with whom to mate in order to produce the offspring of the next generation. Remember, the sex of your mate does not matter, nor does the genotype. You should follow the same mating procedures as you did for the first generation, being sure to record your new genotype after each generation. Class data should be collected after each generation for five generations. At the end of each generation, remember to record the genotype that you have assumed. Your teacher will collect class data after each generation by asking you to raise your hand to report your genotype.

* For simplicity, in this model all individuals in the initial population have a genotype of *Aa*.

4. Allele Frequency: The allele frequencies, p and q, should be calculated for the population after five generations of simulated random mating.

Number of *A* alleles present at the fifth generation

Number of offspring with genotype *AA* ____________ × 2 = __________________ *A* alleles

Number of offspring with genotype *Aa* ____________ × 1 = __________________ *A* alleles

Total = __________________ *A* alleles

$$p = \frac{\text{TOTAL number of } A \text{ alleles}}{\text{TOTAL number of alleles in the population (number of students} \times 2)} = \_\_\_\_\_\_\_\_\_$$

In this case, the total number of alleles in the population is equal to the number of students in the class × 2.

Number of *a* alleles present at the fifth generation

Number of offspring with genotype *aa* ____________ × 2 = __________________ *a* alleles

Number of offspring with genotype *Aa* ____________ × 1 = __________________ *a* alleles

Total = __________________ *a* alleles

$$q = \frac{\text{TOTAL number of } a \text{ alleles}}{\text{TOTAL number of alleles in the population (number of students} \times 2)} = \_\_\_\_\_\_\_\_\_$$

Questions

1. What does the Hardy-Weinberg equation predict for the new p and q?

____

____

2. Do the results you obtained in this simulation agree? If not, why?

____

____

3. What major assumption(s) were not strictly followed in this simulation?

____

____

CASE II — Selection

In this Case you will modify the simulation to make it more realistic. In the natural environment, not all genotypes have the same rate of survival; that is, the environment might favor some genotypes while selecting against others. An example is the human condition of sickle-cell anemia. This is a disease caused by a mutation on one allele, and individuals who are homozygous recessive often do not survive to reach reproductive maturity. For this simulation you will assume that the homozygous recessive individuals never survive (100% selection against), and that heterozygous and homozygous dominant individuals survive 100% of the time.

Procedure

The procedure is similar to that for Case I.

1. Start again with your initial genotype and produce your "offspring" as you did for Case I. This time, however, there is one important difference. Every time your "offspring" is *aa*, it does not reproduce. Since we want to maintain a constant population size, the same two parents must try again until they produce two surviving offspring. You may need to get new "allele" cards from the pool, allowing each individual to complete the activity.

2. Proceed through five generations, selecting against the homozygous recessive offspring 100% of the time. Then add up the genotype frequencies that exist in the population and calculate the new p and q frequencies in the same way you did for Case I.

Questions

1. How do the new frequencies of p and q compare to the initial frequencies in Case I?

___

___

___

2. What major assumption(s) were not strictly followed in this simulation?

___

___

___

3. Predict what would happen to the frequencies of p and q if you simulated another five generations.

___

___

___

4. In a large population would it be possible to completely eliminate a deleterious recessive allele? Explain.

___

___

___

CASE III — Heterozygote Advantage

From Case II it is easy to see what happens to the lethal recessive allele in the population. However, data from many human populations show an unexpectedly high frequency of the sickle-cell allele in some populations. Thus, our simulation does not accurately reflect the real situation; this is because individuals who are heterozygous are slightly more resistant to a deadly form of malaria than homozygous dominant individuals. In other words, there is a slight selection against homozygous dominant individuals as compared to heterozygotes. This fact is easily incorporated into our simulation.

Procedure

1. In this round keep everything the same as it was in Case II, except that if your offspring is *AA*, flip a coin. If the coin lands heads up, the individual does not survive; if tails, the individual does survive.

2. Simulate five generations, starting again with the initial genotype from Case I. The genotype *aa* never survives, and homozygous dominant individuals only survive if the coin toss comes up tails. Since we want to maintain a constant population size, the same two parents must try again until they produce two surviving offspring. Get new "allele" cards from the pool as needed. Total the class genotypes and calculate the p and q frequencies.

3. Starting with the F_5 genotype, go through five more generations, and again total the genotypes and calculate the frequencies of p and q.

4. If time permits, the results from another five generations would be extremely informative.

Questions

1. Explain how the changes in p and q frequencies in Case II compare with Case I and Case III.

____

____

____

2. Do you think the recessive allele will be completely eliminated in either Case II or Case III?

____

____

____

3. What is the importance of heterozygotes (the heterozygote advantage) in maintaining genetic variation in populations?

____

____

____

CASE IV — Genetic Drift (Optional)

It is possible to use our simulation to look at the phenomenon of genetic drift in detail.

Procedure

1. Divide the lab into several smaller "populations" (for example, a class of 30 could be divided into three populations of ten each) so that individuals from one isolated "population" do not interact with individuals from another population.

2. Now go through five generations as you did for Case I. Record the new genotypic frequencies and calculate the new frequencies of p and q for each population.

Questions

1. Explain how the initial genotypic frequencies of the populations compare.

____

____

____

2. What do your results indicate about the importance of population size as an evolutionary force?

____

____

____

Hardy-Weinberg Problems

1. In *Drosophila* the allele for normal-length wings is dominant over the allele for vestigial wings (vestigial wings are stubby little curls that cannot be used for flight). In a population of 1,000 individuals, 360 show the recessive phenotype. How many individuals would you expect to be homozygous dominant and heterozygous for this trait?

2. The allele for unattached earlobes is dominant over the allele for attached earlobes. In a population of 500 individuals, 25% show the recessive phenotype. How many individuals would you expect to be homozygous dominant and heterozygous for this trait?

3. The allele for the hair pattern called "widow's peak" is dominant over the allele for no "widow's peak." In a population of 1,000 individuals, 510 show the dominant phenotype. How many individuals would you expect of each of the possible three genotypes for this trait?

4. In the United States about 16% of the population is Rh negative. The allele for Rh negative is recessive to the allele for Rh positive. If the student population of a high school in the U.S. is 2,000, how many students would you expect for each of the three possible genotypes?

5. In certain African countries 4% of the newborn babies have sickle-cell anemia, which is a recessive trait. Out of a random population of 1,000 newborn babies, how many would you expect for each of the three possible genotypes?

6. In a certain population, the dominant **phenotype** of a certain trait occurs 91% of the time. What is the frequency of the dominant allele?

Data Page

CASE I
Hardy-Weinberg Equilibrium

Initial Class Frequencies:

AA ____ *Aa*____ *aa* ____

My Initial Genotype: ____

F_1 Genotype ____

F_2 Genotype ____

F_3 Genotype ____

F_4 Genotype ____

F_5 Genotype ____

Final Class Frequencies:

AA ____ *Aa*____ *aa* ____

p____ q ____

CASE II
Selection

Initial Class Frequencies:

*AA*____ *Aa*____ *aa* _____

My Initial Genotype: ____

F_1 Genotype ____

F_2 Genotype ____

F_3 Genotype ____

F_4 Genotype ____

F_5 Genotype ____

Final Class Frequencies:

*AA*____ *Aa*____ *aa*____

p____ q ____

CASE III
Heterozygote Advantage

Initial Class Frequencies:

*AA*____ *Aa*____ *aa* _____

My Initial Genotype: ____

F_1 Genotype ____ F_6 Genotype ____

F_2 Genotype ____ F_7 Genotype ____

F_3 Genotype ____ F_8 Genotype ____

F_4 Genotype ____ F_9 Genotype ____

F_5 Genotype ____ F_{10} Genotype ____

Final Class Frequencies: (after five generations)

*AA*____ *Aa*____ *aa* _____

p____ q ____

Final Class Frequencies: (after ten generations)

*AA*____ *Aa*____ *aa* _____

p____ q ____

CASE IV
Genetic Drift

Initial Class Frequencies:

*AA*____ *Aa*____ *aa* _____

p____ q ____

My Initial Genotype: ____

F_1 Genotype ____

F_2 Genotype ____

F_3 Genotype ____

F_4 Genotype ____

F_5 Genotype ____

Final Class Frequencies:

*AA*____ *Aa*____ *aa* _____

p____ q ____

LAB NINE TRANSPIRATION

OVERVIEW

In this lab you will:

1. apply what you learned about water potential from Lab 1 (Diffusion and Osmosis) to the movement of water within the plant,
2. measure transpiration under different lab conditions, and
3. study the organization of the plant stem and leaf as it relates to these processes by observing sections of tissue.

OBJECTIVES

Before doing this lab you should understand:

- how water moves from roots to leaves in terms of the physical/chemical properties of water and the forces provided by differences in water potential;
- the role of transpiration in the transport of water within a plant; and
- the structures used by plants to transport water and regulate water movement.

After doing this lab you should be able to:

- test the effects of environmental variables on rates of transpiration using a controlled experiment, and
- make thin sections of stem, identify xylem and phloem cells, and relate the function of these vascular tissues to the structures of their cells.

INTRODUCTION

The amount of water needed daily by plants for the growth and maintenance of tissues is small in comparison to the amount that is lost through the process of **transpiration** (the evaporation of water from the plant surface) and **guttation** (the loss of liquids from the ends of vascular tissues at the margins of leaves). If this water is not replaced, the plant will wilt and may die.

The transport of water up from the roots in the xylem is governed by differences in **water potential** (the potential energy of water molecules). These differences account for water movement from cell to cell and over long distances in the plant. Gravity, pressure, and solute concentration all contribute to water potential, and water always moves from an area of high water potential to an area of low water potential. The movement itself is facilitated by osmosis, root pressure, and adhesion and cohesion of water molecules.

The Overall Process: Minerals actively transported into the root accumulate in the xylem, increasing solute concentration and decreasing water potential. Water moves in by **osmosis.** As water enters the xylem, it forces fluid up the xylem due to hydrostatic **root pressure.** But this pressure can only move fluid a short distance. The most significant force moving the water and dissolved minerals in the xylem is *upward* pull as a result of **transpiration,** which creates tension. The "pull" on the water from transpiration results from the cohesion and adhesion of water molecules.

The Details: Transpiration begins with evaporation of water through the **stomates** (stomata), small openings in the leaf surface which open into air spaces that surround the mesophyll cells of the leaf. The moist air in these spaces has a higher water potential than the outside air, and water tends to evaporate from the leaf surface (moving from an area of high water potential to an area of lower water potential). The moisture in the air spaces is replaced by water from the adjacent mesophyll cells, lowering their water potential (since the cytoplasm becomes more concentrated). Water will then move into the mesophyll cells by **osmosis** from surrounding cells with higher water potentials, including the xylem. As each water molecule moves into a mesophyll cell, it exerts a pull on the column of water molecules existing in the xylem all the way from the leaves to the roots. This transpirational pull occurs because of (1) the **cohesion** of water molecules to one another due to hydrogen bond formation, and (2) by **adhesion** of water molecules to the walls of the xylem cells which aids in offsetting the downward pull of gravity.

The upward transpirational pull on the fluid in the xylem causes a **tension** (negative pressure) to form in the xylem, pulling the walls of the xylem inward. The tension also contributes to the lowering of water potential in the xylem. This decrease in water potential, transmitted all the way from the leaf to the roots, causes water to move inward from the soil, across the cortex of the root, and into the xylem.

Evaporation through the open stomates is a major route of water loss in plants. However, the stomates must open to allow the entry of CO_2 used in photosynthesis. Therefore, a balance must be maintained between the gain of CO_2 and the loss of water by regulating the opening and closing of stomates on the leaf surface. Many environmental conditions influence the opening and closing of stomates and also affect the rate of transpiration. Temperature, light intensity, air currents, and humidity are some of these factors. Different plants also vary in the rate of transpiration and in the regulation of stomatal opening.

EXERCISE 9A: Transpiration

In this lab you will measure transpiration under various laboratory conditions using a **potometer.** Four suggested plant species are *Impatiens* (which is a moisture-loving plant), *Oleander* (which is more drought tolerant), *Zebrina*, and two-week old *Phaseolus vulgaris* (which are bean seedlings).

Procedure

Each lab group will expose one plant to one treatment.

1. Place the *tip* of a 0.1-mL pipette into a 16-inch piece of clear plastic tubing.

2. Submerge the tubing and pipette in a shallow tray of water. Draw water through the tubing until all bubbles are eliminated.

3. Carefully cut the plant stem under water. This step is very important, because no air bubbles must be introduced into the xylem.

4. While your plant and tubing are submerged, insert the freshly cut stem into the open end of the tubing.

5. Bend the tubing upward into a "U" and use the clamp on a ring stand to hold both the pipette and the tubing (see Figure 9.1).

Figure 9.1

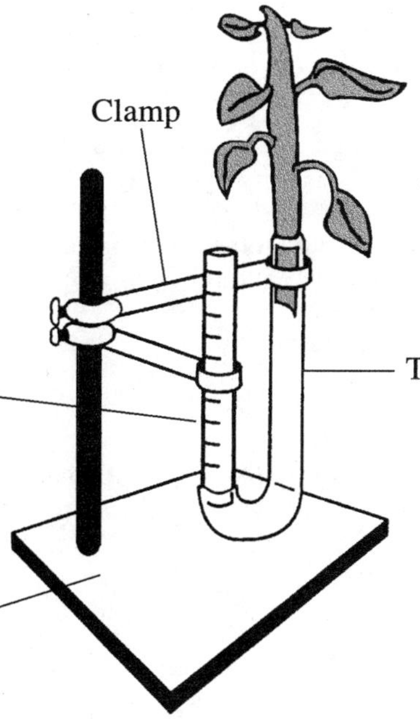

6. If necessary, use petroleum jelly to make an airtight seal surrounding the stem after it has been inserted into the tube. Make sure that the end of the stem is immersed in the water. Do not put petroleum jelly on the cut end of the stem.

7. Let the potometer equilibrate for 10 minutes before recording the time zero reading.

Alternative Procedure for Filling Potometer

(i) Set up the potometer as shown in Figure 9.1.

(ii) Use a water bottle or pipette to fill the tubing. Add water until the water comes out of the tube and no bubbles remain.

(iii) Quickly cut the plant stem and insert it into the potometer.

8. Expose the plant in the tubing to one of the following treatments (you will be assigned a treatment by your teacher):
   - **a.** Room conditions
   - **b.** Floodlight (place a 100-watt bulb 1 meter from the plant and use a beaker filled with water as a heat sink)
   - **c.** Fan (place at least 1 meter from the plant, on low speed, creating a gentle breeze)
   - **d.** Mist (mist the leaves with water and cover with a transparent plastic bag; leave the bottom of the bag open)

9. Read the level of water in the pipette at the beginning of your experiment (time zero) and record your finding in Table 9.1 (page 102).

10. Continue to record the water level in the pipette every 3 minutes for 30 minutes and record the data in Table 9.1.

Table 9.1: Potometer Readings

Time (min)	0	3	6	9	12	15	18	21	24	27	30
Reading (mL)											

11. At the end of your experiment, cut all the leaves off the plant and mass them. Remember to blot off all excess water before massing.

Mass of leaves ________________ grams.

Calculation of Leaf Surface Area

The total surface area of all the leaves can be calculated by using either the Leaf Trace Method or the Leaf Mass Method.

Leaf Trace Method

After arranging all the cut-off leaves on the grid below, trace the edge pattern directly onto Grid 9.1. Count all of the grids that are completely within the tracing and estimate the number of grids that lie partially within the tracing. The grid has been constructed so that a square of 4 blocks equals 1 cm^2. The total surface area can then be calculated by dividing the total number of blocks covered by 4. Record this value here: __________ = Leaf Surface Area (m^2)

Grid 9.1

Leaf Mass Method

- Cut a 1 cm^2 section out of 1 leaf.
- Mass the 1 cm^2 section.
- Multiply the section's mass by 10,000 to calculate the mass per square meter of the leaf. (g/m^2) _____________
- Divide the total mass of the leaves (Step 11) by the mass per square meter (above). This value is the leaf surface area.

$$\frac{\text{Total Mass of Leaves (grams)}}{g/m^2} = \text{Leaf Surface Area } (m^2)$$

- Record this value here: __________ = Leaf Surface Area (m^2)

12. Calculate the water loss per square meter of leaf surface by dividing the water loss at each reading (Table 9.1) by the leaf surface area you calculated. Record your results in Table 9.2.

Table 9.2: Individual Water Loss in mL/m^2

	Time Interval (minutes)										
	0	3	6	9	12	15	18	21	24	27	30
Water Loss (mL)											
Water Loss per m2											

13. Record the averages of the class data for each treatment in Table 9.3.

Table 9.3: Class Average Cumulative Water Loss in mL/m^2

	Time (minutes)										
Treatment	0	3	6	9	12	15	18	21	24	27	30
Room	0										
Light	0										
Fan	0										
Mist	0										

14. For each treatment, graph the average of the class data for each time interval. You may need to convert data to scientific notation. All numbers must be reported to the same power of 10 for graphing purposes.

For this graph you will need to determine the following:

a. The *independent* variable: ____
Use this to label the horizontal (x) axis.

b. The *dependent* variable: ____
Use this to label the vertical (y) axis.

Graph 9.1 **Title:** ____

Analysis of Results

1. Calculate the rate (average amount of water loss per minute per square meter) for each of the treatments.

   Room: ______________________________

   Fan: ______________________________

   Light: ______________________________

   Mist: ______________________________

2. Explain why each of the conditions causes an increase or decrease in transpiration compared with the control.

Condition	Effect	Explanation of Effect
Room		
Fan		
Light		
Mist		

3. Explain the role of water potential in the movement of water from soil through the plant and into the air.

   ____
   ____
   ____
   ____

4. What is the advantage of closed stomata to a plant when water is in short supply? What are the disadvantages?

   ____
   ____
   ____
   ____

5. Describe several adaptations that enable plants to reduce water loss from their leaves. Include both structural and physiological adaptations.

   ____
   ____
   ____
   ____

6. Why did you need to calculate leaf surface area in tabulating your results?

____

____

____

____

EXERCISE 9B: Structure of the Stem

The movement of fluids and nutrients throughout the plant occurs in the vascular tissue: the xylem and phloem of the roots, stems, and leaves. In this exercise you will study the structure of the plant stem by preparing sections of the stem from the plant that you used in Exercise 9A. If your teacher provides you with prepared slides, proceed to Step 15.

Procedure

1. Obtain a nut-and-bolt microtome from your teacher.

2. Turn the nut until it is almost at the end of the bolt, forming a small "cup."

3. Using a new, single-edge razor blade, cut a short piece of plant stem (approximately 5mm — slightly longer than the depth of the "cup" in the nut) from the base of your plant. Make 2 cuts so that both ends are freshly cut. Make sure that this portion of the stem is free of petroleum jelly if you are using the same plant that you used for Exercise 9A.

4. Stand the stem up on its end in the opening of the nut and carefully pour melted paraffin into the nut until it fills the opening, completely covering the stem. Your teacher will direct you in safely melting and pouring the paraffin. (Be careful that the paraffin is not too hot when you pour it or you will cook your stem.) This assembly will allow you to hold your stem upright and cut thin slices.

5. Hold the head of the bolt horizontal on the table with one hand. Holding the razor blade in your other hand, remove the excess wax on top by slicing down to the nut. This technique keeps your fingers out of the way of the razor blade (see Figure 9.2).

Figure 9.2: Using the Nut-and-Bolt Microtome

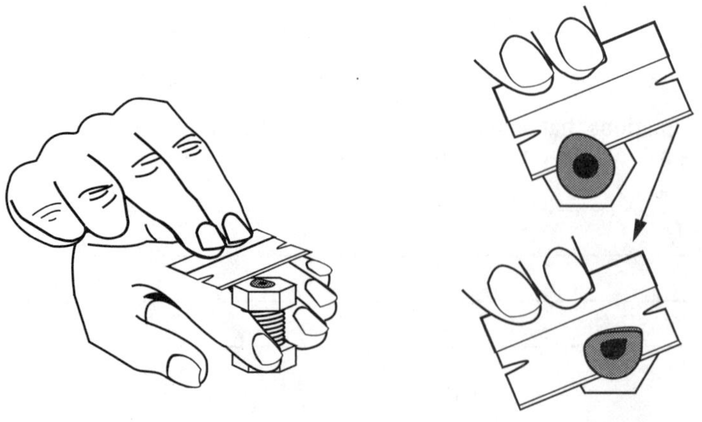

6. Twist the bolt just a little, so a thin core of paraffin and stem sticks up above the surface of the nut.

7. Use a slicing motion to cut this section down to the nut. Use as much of the edge of the razor blade as possible by starting on one end and sliding down to the other with each slice.

8. Put the slice in a dish containing 50% ethanol.

9. Twist the bolt a bit more to get another slice. Remember: you are trying to get the thinnest possible slice. It is better to get part of a thin slice than a thick slice that is entirely round. As you cut each slice, put it in the dish of 50% ethanol. Obtain 8–10 thin sections.

10. Leave the sections in 50% ethanol for 5 minutes. Free the plant tissue from paraffin, if necessary.

11. Using forceps, move the sections to a dish of toluidine blue O stain and leave them there for a short period of time (between 1 and 2 minutes).

12. Rinse the sections in a dish of distilled water.

13. Mount the sections in a drop of 50% glycerin on a microscope slide.

14. Add a cover slip and observe the sections using a compound microscope.

15. Make a drawing of your sections in the space provided in Figure 9.3 (page 108). Identify and label the cell and tissue types described below.

Cell Types

Parenchyma. The most abundant cell type is parenchyma. Parenchyma cells are relatively unspecialized and retain their protoplasts throughout their existence. They have primary cell walls. They make up the mesophyll of leaves (where most of the photosynthetic activity takes place), the flesh of fruits, the pith of stems, and the root and stem cortex. Many parenchyma cells are used for food storage (mainly starch). Starch, you will recall, is a polymer of glucose. Starch forms grains within the parenchyma cells. These grains can be seen inside the cells. Reexamine your section and label the parenchyma cells in your drawing.

Sclerenchyma. Elongated sclerenchyma cells make up fibers and have thick secondary cell walls. They are often lignified, and the protoplasts die at maturity. Fibers may be found in leaves, stems, and fruits. Usually fibers are in bundles, serving a support function, and often are associated with vascular tissue. Check your stem cross section for fibers. They will be found just outside the vascular bundles, their thick walls stained bright blue.

Collenchyma. Many young stems and leaves contain collenchyma cells for support. These cells are living at maturity and characteristically have primary cell walls that are thickened at the corners. Locate collenchyma cells in your cross section.

Tissue Types

Xylem. Xylem is a tissue composed of several different cell types. It is the water-conducting tissue that conveys water and minerals from the soil through the plant. The earliest xylem cells to evolve were fiberlike with thick lignified secondary walls arranged with overlapping ends, with a series of membrane-covered "pits" for passing water from one cell to the next. These are the tracheids. The cells that actually carry the water were misnamed "tracheary elements" in the seventeenth century ("trachea" means air duct) and the name was never corrected. Vessel elements evolved later, first appearing in flowering plants, and are larger in diameter, have holes rather than pits, and offer less resistance to water flow than tracheids. Both vessel elements and tracheids are dead at maturity. Xylem tissue may also contain parenchyma cells and fibers. Look at your cross sections and label the xylem in your drawing.

Phloem. Phloem is a tissue that distributes the carbohydrate products of photosynthesis throughout the plant. This is achieved in flowering plants by the sieve tube members, which have primary cell walls and living protoplasts at maturity but lack nuclei. Companion cells are associated with sieve tube members. These companion cells have nuclei and play an important role in the transfer of substances from cell to cell. Phloem may also contain parenchyma cells and fibers. Look at your cross section. The phloem is located outside the xylem. This aggregation of xylem and phloem is called the vascular bundle. Monocots and dicots have different arrangements of the xylem and phloem tissues, but the cell and tissue types involved are the same.

Epidermis. The epidermis is the outermost layer of cells that serves as a covering for the above-ground plant parts. Some epidermal tissues are covered with a layer of cutin which prevents water loss. The specialized guard cells of the epidermis open and close the stomates. Locate the epidermis on your stem section, then locate guard cells on the leaf section.

Figure 9.3: Stem Cross Section

LAB TEN PHYSIOLOGY OF THE CIRCULATORY SYSTEM

OVERVIEW

1. In Exercise 10A you will learn how to measure blood pressure.
2. In Exercise 10B you will measure pulse rate under different conditions: standing, reclining, after the baroreceptor reflex, and during and immediately after exercise. The blood pressure and pulse rate will be analyzed and related to an index of relative fitness.
3. In Exercise 10C you will measure the effect of temperature on the heart rate of the water flea, *Daphnia magna*.

OBJECTIVES

Before doing this lab you should understand:

- the relationship between temperature and the rates of physiological processes, and
- the basic anatomy of various circulatory systems.

After doing this lab you should be able to:

- measure heart rate and blood pressure in a human volunteer;
- describe the effect of changing body position on heart rate and blood pressure;
- explain how exercise changes heart rate;
- determine a human's fitness index;
- analyze cardiovascular data collected by the entire class; and
- discuss and explain the relationship between heart rate and temperature.

INTRODUCTION

The cardiovascular (circulatory) system functions to deliver oxygen and nutrients to tissues for growth and metabolism, and to remove metabolic wastes. The heart pumps blood through a circuit that includes arteries, arterioles, capillaries, venules, and veins. One important circuit is the pulmonary circuit, where there is an exchange of gases within the alveoli of the lungs. The right side of the human heart receives deoxygenated blood from body tissues and pumps it to the lungs. The left side of the heart receives oxygenated blood from the lungs and pumps it to the tissues.

With increased exercise, several changes occur within the circulatory system, thus increasing the delivery of oxygen to actively respiring muscle cells. These changes include increased heart rate, increased blood flow to muscular tissue, decreased blood flow to nonmuscular tissue, increased arterial pressure, increased body temperature, and increased breathing rate.

Blood Pressure

An important measurable aspect of the circulatory system is blood pressure. When the ventricles of the heart contract, pressure is increased throughout all the arteries. Arterial blood pressure is directly dependent on the amount of blood pumped by the heart per minute and the resistance to blood flow through the arterioles. The arterial blood pressure is determined using a device known as a **sphygmomanometer.** This device consists of an inflatable cuff connected by rubber hoses to a hand pump and to a pressure gauge graduated in millimeters of mercury. The cuff is wrapped around the upper arm and inflated to a pressure that will shut off the brachial artery. The examiner listens for the sounds of blood flow in the brachial artery by placing the bell of a stethoscope in the inside of the elbow below the biceps (Figure 10.1).

Figure 10.1: The Use of a Sphygmomanometer to Measure Blood Pressure

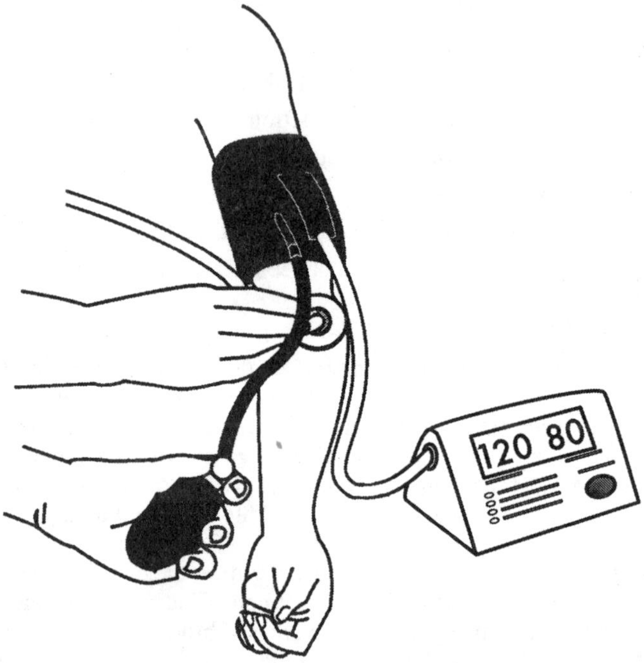

At rest, the blood normally goes through the arteries so that the blood in the central part of the artery moves faster than the blood in the peripheral part. Under these conditions, the artery is silent when one listens. When the sphygmomanometer cuff is inflated to a pressure above the systolic pressure, the flow of blood is stopped and the artery is again silent. As the pressure in the cuff gradually drops to levels between the systolic and diastolic pressures of the artery, the blood is pushed through the compressed walls of the artery in a turbulent flow. Under these conditions, the blood is mixed, and the turbulence sets up vibrations in the artery that are heard as sounds in the stethoscope. These sounds are known as the heart sounds, or sounds of Korotkoff.

The sounds are divided into five phases based on the loudness and quality of the sounds.

Phase 1. A loud, clear, snapping sound is evident which increases in intensity as the cuff is deflated. In the example shown in Figure 10.2, this phase begins at a cuff pressure of 120 millimeters of mercury (mm Hg) and ends at a pressure of 106 mm Hg.

Phase 2. A succession of murmurs can be heard. Sometimes the sounds seem to disappear during this time, which may be a result of inflating or deflating the cuff too slowly. In the example shown in Figure 10.2, this phase begins at a cuff pressure of 106 mm Hg and ends at a pressure of 86 mm Hg.

Phase 3. A loud, thumping sound, similar to that in Phase 1, but less clear, replaces the murmurs. In the example shown in Figure 10.2, Phase 3 begins at a cuff pressure of 86 mm Hg and ends at a pressure of 81 mm Hg.

Phase 4. A muffled sound abruptly replaces the thumping sounds of Phase 3. In the example shown in Figure 10.2, this phase begins at a cuff pressure of 81 mm Hg and ends at a pressure of 76 mm Hg.

Phase 5. All sounds disappear.

Figure 10.2: The Five Phases of the Heart Sounds (Sounds of Korotkoff)

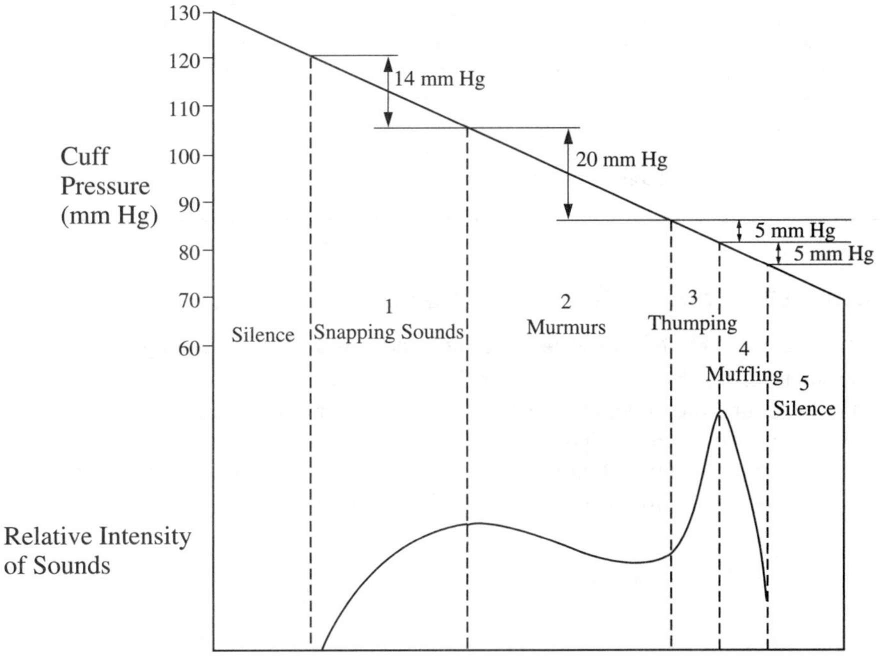

The cuff pressure at which the first sound is heard (that is, the beginning of Phase 1) is taken as the systolic pressure. The cuff pressure at which the muffled sound of Phase 4 disappears (the beginning of Phase 5) is taken as the measurement of the diastolic pressure. In the example shown in Figure 10.2, the pressure would be recorded in this example as 120/76. A normal blood pressure measurement for a given individual depends on the person's age, sex, heredity, and environment. When these factors are taken into account, blood pressure measurements that are chronically elevated may indicate a state deleterious to the health of the person. This condition is called hypertension and is a major contributing factor in heart disease and stroke. Typical blood pressure for men and women varies with age and fitness (Table 10.1). For high school students, the typical range is usually 100–120/70–90.

Table 10.1: Typical Blood Pressure for Men and Women at Different Ages

	Systolic		Diastolic	
Age (in years)	Men	Women	Men	Women
10	103	103	69	70
11	104	104	70	71
12	106	106	71	72
13	108	108	72	73
14	110	110	73	74
15	112	112	75	76
16	118	116	73	72
17	121	116	74	72
18	120	116	74	72
19	122	115	75	71
20–24	123	116	76	72
25–29	125	117	78	74
30–34	126	120	79	75
35–39	127	124	80	78
40–44	129	127	81	80
45–49	130	131	82	82
50–54	135	137	83	84
55–59	138	139	84	84
60–64	142	144	85	85
65–69	143	154	83	85
70–74	145	159	82	85

EXERCISE 10A: Measuring Blood Pressure

Note: These labs are ONLY for experimental, and not diagnostic, purposes.

A sphygmomanometer (blood pressure cuff) is used to measure blood pressure. The cuff, designed to fit around the upper arm, can be expanded by pumping a rubber bulb connected to the cuff. The pressure gauge, scaled in millimeters, indicates the pressure inside the cuff. A stethoscope is used to listen to the individual's pulse (see Figure 10.1). The earpieces of the stethoscope should be cleaned with alcohol swabs before and after each use.

Procedure

1. Work in pairs. Those who are to have their blood pressure measured should be seated with both shirt sleeves rolled up.

2. Attach the cuff of the sphygmomanometer snugly around the upper arm.

3. Place the stethoscope directly below the cuff in the bend of the elbow joint.

4. Close the valve of the bulb by turning it clockwise. Pump air into the cuff until the pressure gauge reaches 180 mm Hg.

5. Turn the valve of the bulb counterclockwise and slowly release air from the cuff. Listen for a pulse.

6. When you first hear the heart sounds, note the pressure on the gauge. *This is the systolic pressure.*

7. Continue to slowly release air and listen until the clear thumping sound of the pulse becomes strong and then fades. When you last hear the full heart beat, note the pressure. *This is the diastolic pressure.*

8. Repeat the measurement two more times and determine the average systolic and diastolic pressure, then record these values in the blood pressure data box on page 118.

9. Trade places with your partner. When your average systolic and diastolic pressure have been determined, record these values in the blood pressure data box.

EXERCISE 10B: A Test of Fitness

The point scores on the following tests provide an evaluation of fitness based not only on cardiac muscular development but also on the ability of the cardiovascular system to respond to sudden changes in demand. **CAUTION: Make sure that you do not attempt this exercise if strenuous activity will aggravate a health problem.**

Work in pairs. Determine the fitness level for one member of the pair (Tests 1 to 5) and then repeat the process for the other member.

Test 1: Standing Systolic Compared with Reclining Systolic

Use the sphygmomanometer as you did in Exercise 10A to measure the change in systolic blood pressure from a reclining to a standing position.

Procedure

1. The subject should recline on a lab bench for at least 5 minutes. At the end of this time, measure the systolic and diastolic pressure and record these values below.

reclining systolic pressure _____ mm Hg   reclining diastolic pressure _____ mm Hg

2. Remain reclining for two minutes, then stand and *immediately* repeat measurements on the same subject (arms down). Record these values below.

   standing systolic pressure _____ mm Hg   standing diastolic pressure _____ mm Hg

3. Determine the change in systolic pressure from reclining to standing by subtracting the standing measurement from the reclining measurement. Assign fitness points based on Table 10.2 and record in the fitness data box on page 118.

Table 10.2: Change in Systolic Pressure from Reclining to Standing

Change (mm Hg)	Fitness Points
rise of 8 or more	3
rise of 2–7	2
no rise	1
fall of 2–5	0
fall of 6 or more	-1

Cardiac Rate and Physical Fitness

During physical exertion, the cardiac rate (beats per minute) increases. This increase can be measured as an increase in pulse rate. Although the maximum cardiac rate is generally the same in people of the same age group, those who are physically fit have a higher stroke volume (milliliters per beat) than more sedentary individuals. A person who is in poor physical condition, therefore, reaches his or her maximum cardiac rate at a lower work level than a person of comparable age who is in better shape. Individuals who are in good physical condition can deliver more oxygen to their muscles (have a higher aerobic capacity) before reaching maximum cardiac rate than can those in poor condition.

Thus, the physically fit have a slower increase in their cardiac rate with exercise and a faster return to the resting cardiac rate after exercise. Physical fitness, therefore, involves not only muscular development but also the ability of the cardiovascular system to respond to sudden changes in demand.

Test 2: Standing Pulse Rate

Procedure

1. The subject should stand at ease for 2 minutes after Test 1.
2. After the 2 minutes, determine the subject's pulse.
3. Count the number of beats for 30 seconds and multiply by 2. The pulse rate is the number of beats per minute. Record them in the fitness data box. Assign fitness points based on Table 10.3 and record them in the fitness data box.

Table 10.3: Standing Pulse Rate

Pulse Rate (beats/min)	Fitness Points
60–70	3
71–80	3
81–90	2
91–100	1
101–110	1
111–120	0
121–130	0
131–140	-1

Test 3: Reclining Pulse Rate

Procedure

1. The subject should recline for 5 minutes on a lab bench.
2. Determine the subject's resting pulse rate.
3. Count the number of beats for 30 seconds and multiply by 2. (**Note:** the subject should remain reclining for the next test.) The pulse rate is equal to the number of beats per minute. Record it in the fitness data box. Assign fitness points based on Table 10.4 and record them in the fitness data box.

Table 10.4: Reclining Pulse Rate

Pulse Rate (beats/min)	Fitness Points
50–60	3
61–70	3
71–80	2
81–90	1
91–100	0
101–110	-1

Test 4: Baroreceptor Reflex (Pulse Rate Increase from Reclining to Standing)

Procedure

1. The reclining subject should now stand up.
2. *Immediately* take the subject's pulse by counting the number of beats for 30 seconds. Multiply by 2 to determine the pulse rate in beats per minute. Record this value below. The observed increase in pulse rate is initiated by baroreceptors (pressure receptors) in the carotid artery and in the aortic arch. When the baroreceptors detect a drop in blood pressure they signal the medulla of the brain to increase the heartbeat and, consequently, the pulse rate.

   Pulse immediately upon standing = _______ beats per min
3. Subtract the reclining pulse rate (recorded in Test 3) from the pulse rate immediately upon standing (recorded in Test 4) to determine the pulse rate increase upon standing. Record in the fitness data box. Assign fitness points based on Table 10.5 and record in the fitness data box.

Table 10.5: Pulse Rate Increase from Reclining to Standing

Reclining Pulse	Pulse Rate Increase on Standing (# beats)				
(beats/min)	0–10	11–18	19–26	27–34	35–43
	Fitness Points				
50–60	3	3	2	1	0
61–70	3	2	1	0	-1
71–80	3	2	0	-1	-2
81–90	2	1	-1	-2	-3
91–100	1	0	-2	-3	-3
101–110	0	-1	-3	-3	-3

Test 5: Step Test — Endurance

Procedure

1. The subject should do the following: Place your right foot on an 18-inch high stool. Raise your body so that your left foot comes to rest by your right foot. Return your left foot to the original position. Repeat this exercise 5 times, allowing 3 seconds for each step up.
2. *Immediately* after the completion of this exercise, measure the subject's pulse for 15 seconds and record below; measure again for 15 seconds and record; continue taking the subject's pulse and recording the rates at 60, 90, and 120 seconds.

   Number of beats in the 0- to 15-second interval ____ $\times$ 4 = ____ beats per min
   Number of beats in the 16- to 30-second interval ____ $\times$ 4 = ____ beats per min
   Number of beats in the 31- to 60-second interval ____ $\times$ 2 = ____ beats per min
   Number of beats in the 61- to 90-second interval ____ $\times$ 2 = ____ beats per min
   Number of beats in the 91- to 120-second interval ____ $\times$ 2 = ____ beats per min

3. Observe the time that it takes for the subject's pulse rate to return to approximately the level that was recorded in Test 2. Assign fitness points based on Table 10.6 and record them in the fitness data box.

Table 10.6: Time Required for Return of Pulse Rate to Standing Level after Exercise

Time (seconds)	Fitness Points
0–30	4
31–60	3
61–90	2
91–120	1
121+	1
1–10 beats above standing pulse rate	0
11–30 beats above standing pulse rate	-1

4. Subtract the subject's normal standing pulse rate (recorded in Test 2) from his/her pulse rate immediately after exercise (the 0- to 15-second interval) to obtain pulse rate increase. Record this on the data sheet. Assign fitness points based on Table 10.7 and record them in the fitness data box.

Table 10.7: Pulse Rate Increase after Exercise

Standing Pulse (beats/min)	Pulse Rate Increase Immediately after Exercise (# beats)				
	0–10	11–20	21–30	31–40	41+
	Fitness Points				
60–70	3	3	2	1	0
71–80	3	2	1	0	-1
81–90	3	2	1	-1	-2
91–100	2	1	0	-2	-3
101–110	1	0	-1	-3	-3
111–120	1	-1	-2	-3	-3
121–130	0	-2	-3	-3	-3
131–140	0	-3	-3	-3	-3

DATA SHEET

Blood Pressure Data

Measurement	1	2	3	Average
Systolic				
Diastolic				

Fitness Data

	Measurement	Points
Test 1. Change in systolic pressure from reclining to standing	mm Hg	
Test 2. Standing pulse rate	beats / min	
Test 3. Reclining pulse rate	beats / min	
Test 4. Baroreceptor reflex Pulse rate increase on standing	beats / min	
Test 5. Step Test Return of pulse to standing rate after exercise Pulse rate increase immediately after exercise	seconds beats / min	
	TOTAL SCORE	

Total Score	**Relative Cardiac Fitness**
18–17	Excellent
16–14	Good
13–8	Fair
7 or less	Poor

Topics for Discussion

1. Explain why blood pressure and heart rate differ when measured in a reclining position and in a standing position.

2. Explain why high blood pressure is a health concern.

3. Explain why an athlete must exercise harder or longer to achieve a maximum heart rate than a person who is not as physically fit.

4. Research and explain why smoking causes a rise in blood pressure.

____

____

____

____

____

____

____

____

EXERCISE 10C: Heart Rate and Temperature

In ectothermic animals there is a direct relationship between the rate of many physiological activities and environmental temperature. The rate of metabolism in these animals increases as environmental temperatures increase from approximately 5°C to 35°C. Increasing the temperature by approximately 10°C results in a doubling of the metabolic rate. That is why a snake or lizard can hardly move when it is cold but becomes quite active after warming in the sun.

Procedure

1. Pick up a *Daphnia* with a large-bore pipette or eye dropper (a broken-off Pasteur pipette will also work).

2. Place the *Daphnia* into the large end of a Pasteur pipette and allow the culture fluid containing the *Daphnia* to run down into the narrow tip of the pipette.

3. Use a paper towel to draw some of the culture fluid out of the pipette until the *Daphnia* no longer moves down the tube and the fluid level is approximately 5 mm above the *Daphnia*.

4. Seal the narrow end of the pipette with clay or petroleum jelly.

5. Score the pipette with a file and break it off about 2 cm above the *Daphnia*. Seal the broken end by keeping the pipette upright (sealed end down) and inserting the broken end into clay or petroleum jelly.

6. Place the tube containing the *Daphnia* into a petri dish or bowl of water that is the same temperature as the culture fluid. Use a dissecting microscope to observe the *Daphnia*. Refer to Figure 10.4 (page 122) to locate the *Daphnia*'s heart. Count the heartbeats for 10 seconds, then multiply by 6 to obtain the heart rate in beats per minute. Record the temperature and heart rate in Table 10.8.

7. Now place the tube into a petri dish containing water at 10 to 15°C. Note the temperature and the heart rate after the rate has stabilized and record it in Table 10.8.

8. Slowly add warm water (not greater than 35°C) to the dish. In Table 10.8 record the temperature and changes in heart rate for every 5°C change in temperature until you can no longer accurately count the beats.

Alternative Procedure

(i) Obtain two concave depression slides. Pull off several cotton fibers from a cotton ball and place these in the depression of 1 slide.

(ii) Add a *Daphnia* to the slide. Place the second slide on top, concave side over the *Daphnia*, and secure the 2 slides with 2 rubber bands. Leave 1 strand of rubber band between the slides to hold them apart for sufficient circulation (see Figure 10.3).

Figure 10.3

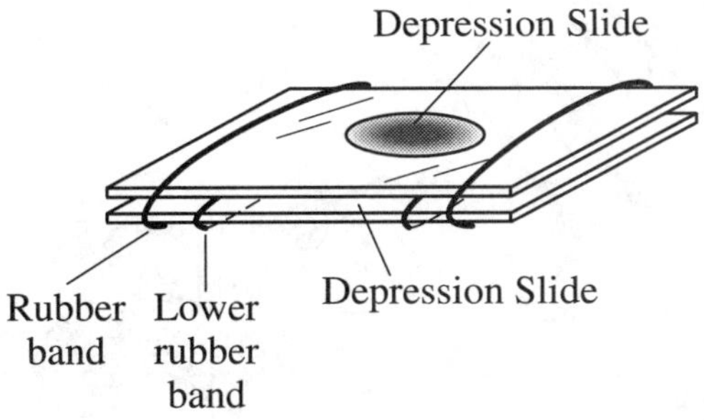

(iii) Use several culture dishes to set up baths of varying temperatures. Begin by placing the slide into the coolest bath.

(iv) Use a dissecting microscope to observe the *Daphnia*. Refer to Figure 10.4 to locate the *Daphnia*'s heart. Count the heartbeats for 10 seconds, then multiply by 6 to obtain the heart rate per minute.

(v) Now place the slides into the next warmest bath. Record the temperature and the heart rate after the rate has stabilized.

(vi) In Table 10.8 record the temperature and changes in heart rate for every change in temperature until you can no longer accurately count the beats.

Table 10.8: Temperature and Heart Rate Data

Reading	Temperature (°C)	Heart Rate (beats/min)
1		
2		
3		
4		
5		
6		
7		
8		

Figure 10.4: *Daphnia*

(Note the position of the heart.)

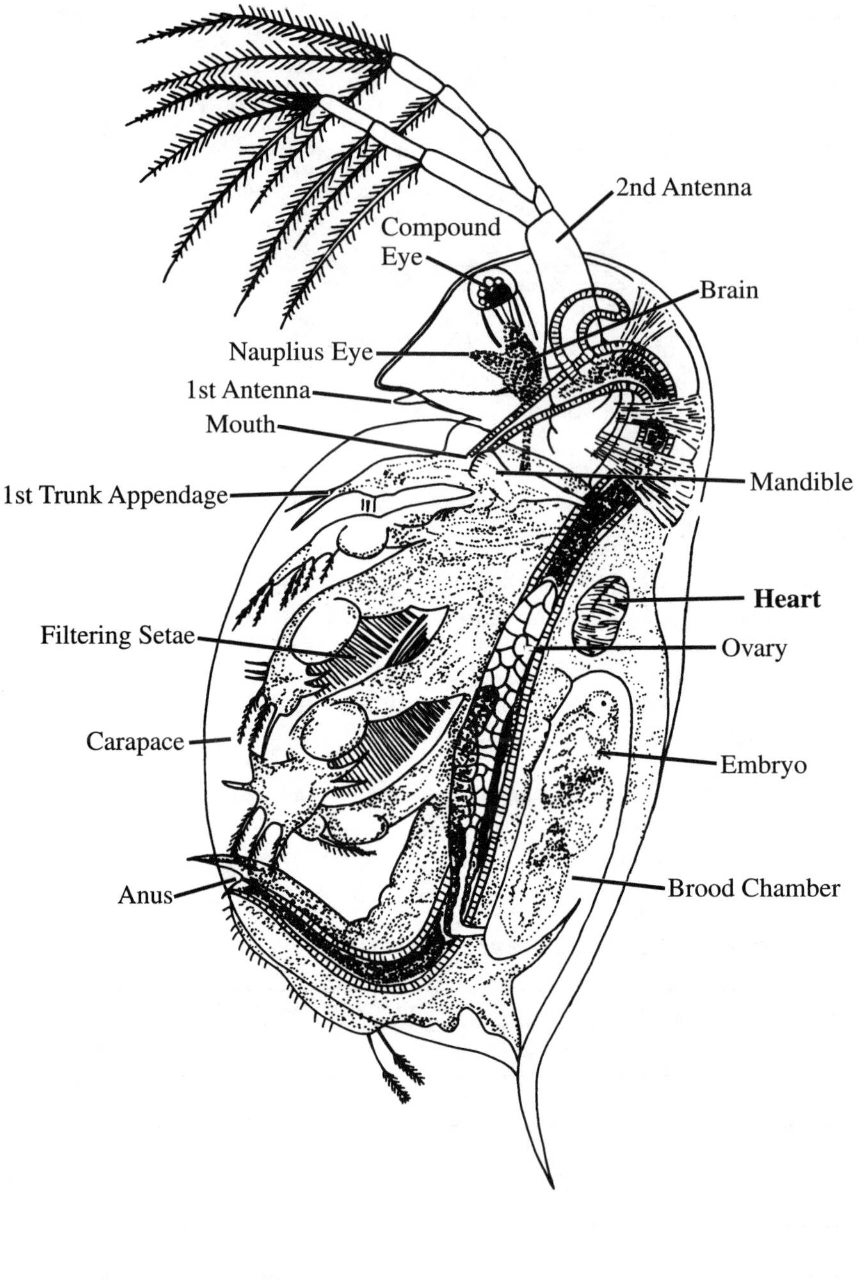

Analysis of Results

Graph the temperature and heart rate data. For this graph you will need to determine the following:

a. The *independent* variable: ____
Use this to label the horizontal (x) axis.

b. The *dependent* variable: ____
Use this to label the vertical (y) axis.

Graph 10.1 **Title:** ____

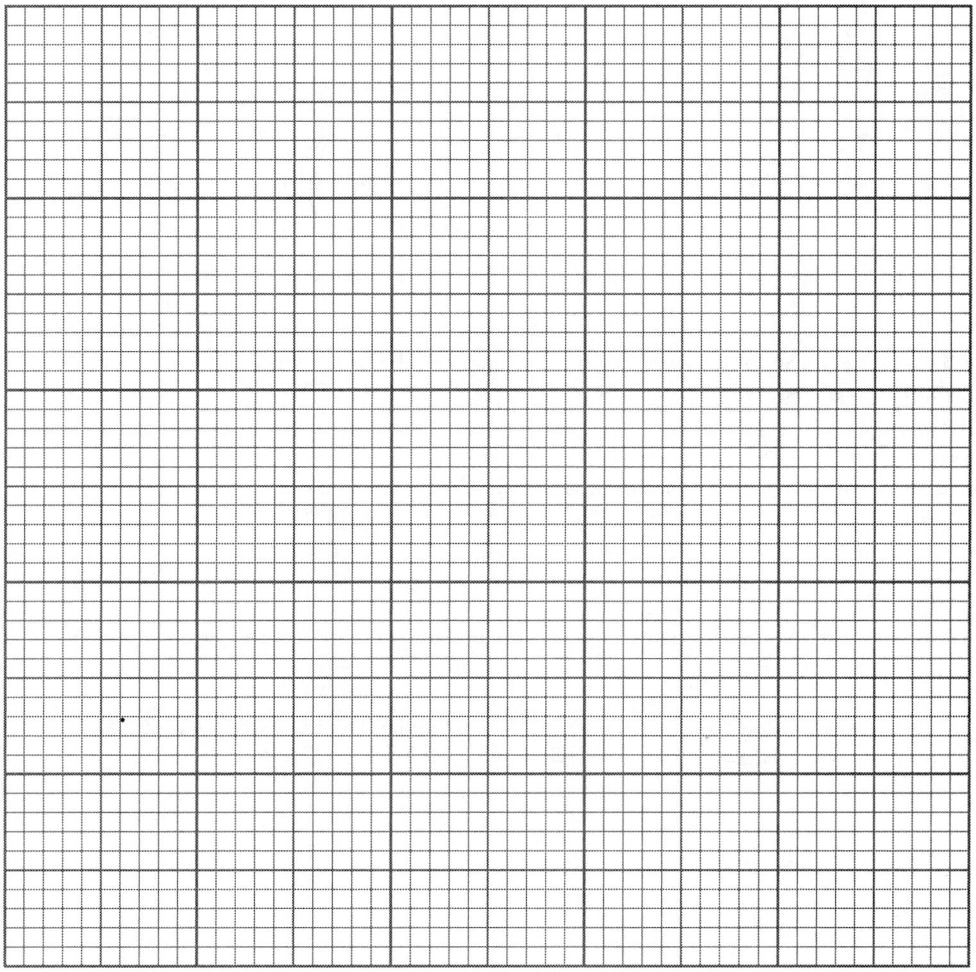

Topics for Discussion

1. Why does temperature affect heart rate in ectothermic organisms?

2. Discuss what results you might obtain if you repeated this experiment using an endothermic organism.

3. Describe at least four ways an ectothermic organism's behavior helps it regulate its temperature.

For further information see *Laboratory Guide to Human Physiology: Concepts and Clinical Applications*, 4th ed., by Stuart J. Fox (Dubuque, IA: William C. Brown Publishing Company, 1987).

LAB ELEVEN ANIMAL BEHAVIOR

OVERVIEW

In this lab you will observe some aspects of animal behavior.

1. In Exercise 11A you will observe pillbugs and design an experiment to investigate their responses to environmental variables.
2. In Exercise 11B you will observe and investigate mating behavior in fruit flies. Your teacher may suggest other organisms or other types of animal behavior to study.

OBJECTIVES

Before doing this lab you should understand:

- the concept of distribution of organisms in a resource gradient, and
- the difference between kinesis and taxis.

After doing this lab you should be able to:

- describe some aspects of animal behavior, such as orientation behavior, agonistic behavior, dominance display, or mating behavior, and
- understand the adaptiveness of the behaviors you studied.

INTRODUCTION

Ethology is the study of animal behavior. Behavior is an animal's response to sensory input and falls into two basic categories: **learned** and **innate** (inherited).

Orientation behaviors place the animal in its most favorable environment. In **taxis** the animal moves toward or away from a stimulus. Taxis is often exhibited when the stimulus is light, heat, moisture, sound, or chemicals. **Kinesis** is a movement that is random and does not result in orientation with respect to a stimulus. If an organism responds to bright light by moving away, that is taxis. If an animal responds to bright light by random movements in all directions, that is kinesis.

Agonistic behavior is exhibited when animals respond to each other by aggressive or submissive responses. Often the agonistic behavior is simply a display that makes the organism look big or threatening. It is sometimes studied in the laboratory with *Bettas* (Siamese Fighting Fish).

Mating behaviors may involve a complex series of activities that facilitate finding, courting, and mating with a member of the same species.

EXERCISE 11A: General Observation of Behaviors

In this lab you will be working with terrestrial isopods commonly known as pillbugs, sowbugs, or roly-polies. These organisms are members of the Phylum *Arthropoda*, Class *Crustacea*, which also includes shrimp and crabs. Most members of this group respire through gills.

Procedure

1. Place 10 pillbugs and a small amount of bedding material in a petri dish. Pillbugs generally do not climb, but if they do, you may cover the dish with plastic wrap or the petri dish cover.

2. Observe the pillbugs for 10 minutes. Make notes on their general appearance, movements about the dish, and interactions with each other. Notice if they seem to prefer one area over another, if they keep moving, settle down, or move sporadically. Note any behaviors that involve 2 or more pillbugs. Try to make your observations without disturbing the animals in any way. Do not prod or poke or shake the dish, make loud sounds, or subject them to bright lights. You want to observe their behavior, not influence it or interfere with it.

3. Make a detailed sketch of a pillbug.

Kinesis in Pillbugs

1. Prepare a choice chamber as illustrated in Figure 11.1. The choice chamber consists of two large, plastic petri dishes taped together with an opening cut between them. Cut the opening with scissors and use tape to hold the dishes together. Line one chamber with a moist piece of filter paper and the other with a dry piece of filter paper.
2. Use a soft brush to transfer 10 pillbugs from the stock culture into the choice chamber. Place 5 pillbugs in each side of the choice chamber. Cover the chambers.
3. Count how many pillbugs are on each side of the choice chamber every 30 seconds for 10 minutes and then record your data in Table 11.1 (page 128). Continue to record even if they all move to one side or stop moving.
4. Return your pillbugs to the stock culture.
5. Graph both the number of pillbugs in the wet chamber and the number in the dry chamber using Graph 11.1 on page 129.

Figure 11.1 Choice Chamber

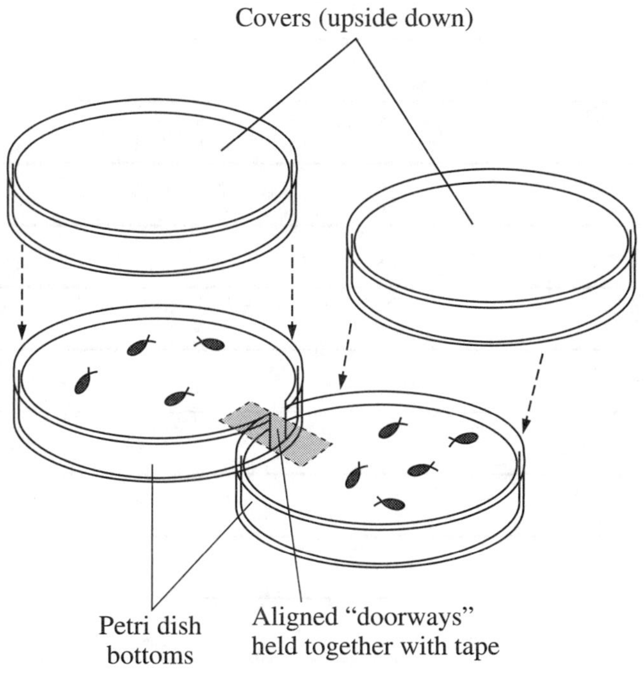

Table 11.1

Time (mins.)	Number in Wet Chamber	Number in Dry Chamber	Other Notes
0			
0.5			
1.0			
1.5			
2.0			
2.5			
3.0			
3.5			
4.0			
4.5			
5.0			
5.5			
6.0			
6.5			
7.0			
7.5			
8.0			
8.5			
9.0			
9.5			
10.0			

For this graph you will need to determine the following:

a. The *independent* variable: ______________________________
Use this to label the horizontal (x) axis.

b. The *dependent* variable: ______________________________
Use this to label the vertical (y) axis.

Graph 11.1 **Title:** ______________________________

Analysis

1. What conclusions do you draw from your data? Explain the physiological reasons for the behavior observed in this activity.

____

____

____

____

____

2. Obtain results from all of the lab groups in your class. With respect to humidity, light, temperature, and other environmental conditions, which types of environment do isopods prefer? How do the data support these conclusions? Give specific examples.

____

____

____

____

____

3. How do isopods locate appropriate environments?

____

____

4. If you suddenly turned a rock over and found isopods under it, what would you expect them to be doing? If you watched the isopods for a few minutes, how would you expect to see their behavior change?

____

____

____

____

____

5. Is the isopod's response to moisture best classified as kinesis or taxis? Explain your response.

____

____

____

____

____

Student-Designed Experiment to Investigate Pillbugs' Response to Temperature, pH, Background Color, Light, or Another Variable

1. Select one of the variable factors listed above and develop a hypothesis concerning the pillbugs' response to the factor.

____

____

____

2. Use the material available in your classroom to design an experiment (see Appendix I for information on how to design experiments). Remember that heat is generated by lamps.

   a. State the objective of your experiment.

   ____

   ____

   ____

   b. List the materials you will use.

   ____

   ____

   ____

   c. Outline your procedure in detail.

   ____

   ____

   ____

   ____

   ____

   ____

   ____

3. Decide what data you will collect and design your data sheet.

4. Run your experiment.

5. Make any graphical representation of your data that will help to visualize or interpret the data.

6. Write a conclusion based on your experimental results.

7. Return your isopods to the stock culture.

EXERCISE 11B: Reproductive Behavior in Fruit Flies

Introduction

In this experiment, you will place 3 or 4 virgin female *Drosophila melanogaster* flies in the same vial with 3 or 4 male flies and, using a dissecting microscope or hand lens, observe the behavior of each sex. Mating in *Drosophila melanogaster* follows a strict behavioral pattern. Five phases can be distinguished (see Figure 11.2):

A. Orientation
B. Male song (wing vibration)
C. Licking of the female genitalia
D. Attempted copulation
E. Copulation
F. Rejection (extrusion of ovipositor)

Figure 11.2: Courtship Behavior in *Drosophila melanogaster*

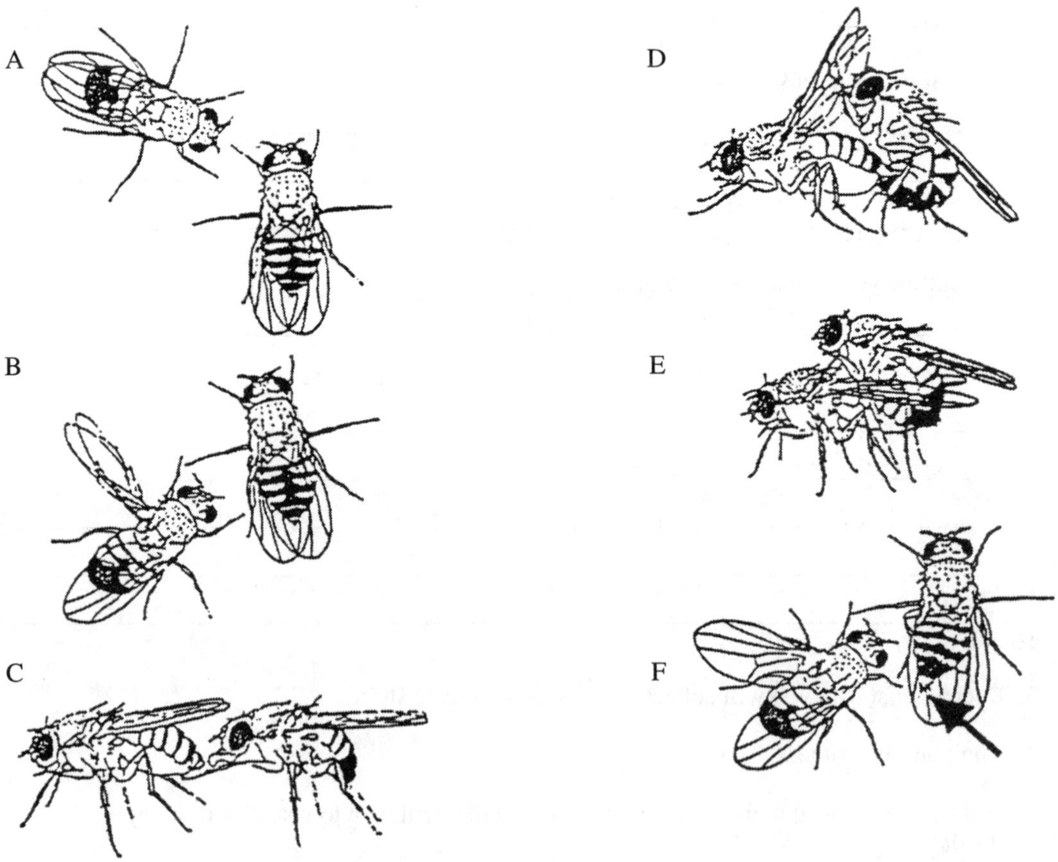

Fig. 11.2 A–F. Courtship behavior in *D. melanogaster*. A, orientation of the male towards the female. B, wing vibration by the male. C, the male licks the female's genitalia with his proboscis. D, mounting by the male with genital contact. E, flies in copulation. F, a rejection response by the female. The female turns her abdomen towards the male and extends her ovipositor (see arrow).

At least 14 different behaviors have been described. Listed below are 10 of the most common and easily recognized of these behaviors. Read the descriptions carefully so that you will be familiar with the behaviors. Six of the behaviors are seen in males, 4 in females.

Male Behaviors

1. **Wing vibration.** The male extends one or both wings from the resting position and moves them rapidly up and down.
2. **Waving.** The wing is extended and held 90° from the body, then relaxed without vibration.
3. **Tapping.** The forelegs are extended to strike or tap the female.
4. **Licking.** The male licks the female's genitalia (on the rear of her abdomen).
5. **Circling.** The male postures and then circles the female, usually when she is nonreceptive.
6. **Stamping.** The male stamps forefeet as in tapping but does not strike the female.

Female Behaviors

1. **Extruding.** A temporary tubelike structure is extended from the female's genitalia (see Figure 11.2F).
2. **Decamping.** A nonreceptive female runs, jumps, or flies away from the courting male.
3. **Depressing.** A nonreceptive female prevents access to her genitalia by depressing her wings and curling the tip of her abdomen down.
4. **Ignoring.** A nonreceptive female ignores the male.

Procedure

1. Set up the stereoscopic microscope.
2. Have paper and pencil ready. The behaviors may happen very rapidly. One person should call out observations while the other person records.
3. Obtain one vial containing virgin females and one vial containing males, and gently tap the male flies into the female vial.
4. Observe first with the naked eye and, once the flies have encountered each other, use the stereoscopic microscope to make observations.
5. As you identify the various behaviors, record their sequence and duration. Quantify your observations. To do this you may consider counting the number of times a behavior takes place and timing the duration of behaviors.
6. Discuss possible original experiments investigating reproductive behavior in flies.

Results

Prepare a detailed account of the behaviors you have observed. Include sketches and quantitative analysis as appropriate.

Student-Designed Experiment to Investigate Reproductive Behavior in Fruit Flies

Design a simple experiment to investigate one of the following questions or any other that you devise:

a. Will males placed in a vial with only males demonstrate courtship behaviors?
b. Will males respond to dead females?
c. Do males compete?
d. How will males respond to already mated females?

1. Develop a hypothesis concerning the fruit fly behavior.

____

____

____

2. State your objective.

____

____

____

3. List the materials you will use.

____

____

____

4. Outline your procedure in detail.

____

____

____

____

____

____

____

____

____

____

____

____

5. Decide what data you will collect and design your data sheet.

6. Run your experiment.

7. Make any graphical representation of your data that will help to visualize or interpret the data.

8. Write a conclusion based on your experimental results.

LAB TWELVE DISSOLVED OXYGEN AND AQUATIC PRIMARY PRODUCTIVITY

OVERVIEW

1. In Exercise 12A you will measure and analyze the dissolved oxygen (DO) concentration in water samples at varying temperatures.

2. In Exercise 12B you will measure and analyze the primary productivity of natural waters or lab cultures using screens to simulate the attenuation (decrease) of light with increasing depth.

OBJECTIVES

Before doing this lab you should understand:

- the biological importance of carbon and oxygen cycling in ecosystems,
- how primary productivity relates to the metabolism of organisms in an ecosystem,
- the physical and biological factors that affect the solubility of gases in aquatic ecosystems, and
- the relationship between dissolved oxygen and the processes of photosynthesis and respiration and how these processes affect primary productivity.

After doing this lab you should be able to:

- measure primary productivity based on changes in dissolved oxygen in a controlled experiment, and
- investigate the effects of changing light intensity on primary productivity in a controlled experiment.

INTRODUCTION

In the aquatic environment, oxygen must be in solution in a free state (O_2) before it is available for use by organisms. Its concentration and distribution in the aquatic environment are directly dependent on chemical and physical factors and are greatly affected by biological processes. In the atmosphere there is an abundance of oxygen, with about 200 milliliters of oxygen for every liter of air. Conversely, in the aquatic environment there are only about 5 to 10 milliliters of dissolved oxygen in a liter of water. The concentration of the oxygen in aquatic environments is a very important component of water quality.

At 20°C oxygen diffuses 300,000 times faster in air than in water, making the distribution of oxygen in air relatively uniform. Spatial distribution of oxygen in water, on the other hand, can be highly variable, especially in the absence of mixing by currents, winds, or tides.

Other chemical and physical factors, such as salinity, pH, and especially temperature, can affect the DO concentration and distribution. Salinity, usually expressed in parts per thousand (ppt), is the content of dissolved salts in water. Generally, as temperature and salinity increase, the solubility of oxygen in water decreases (Figure 12.1).

Figure 12.1: Solubility of Oxygen in Water

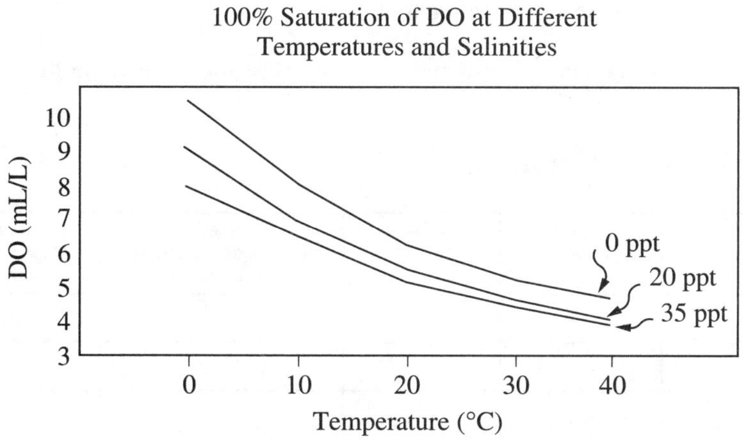

The partial pressure of oxygen in the air above the water affects the amount of DO in the water. Less oxygen is present at higher elevations since the air itself is less dense; therefore, water at higher elevations contains less oxygen. At 4,000 meters in elevation (about 13,000 feet), the amount of dissolved oxygen in water is less than two-thirds what it is at sea level. All of these physical factors work together to increase diversity in aquatic habitats with regard to oxygen availability.

Biological processes, such as photosynthesis and respiration, can also significantly affect DO concentration. Photosynthesis usually increases the DO concentration in water. Aerobic respiration requires oxygen and will usually decrease DO concentration. The measurement of the DO concentration of a body of water is often used to determine whether the biological activities requiring oxygen are occurring; consequently, it is an important indicator of pollution.

EXERCISE 12A: Dissolved Oxygen and Temperature

There are several brands of test kits available to determine the dissolved oxygen content of a water sample. Follow your teacher's instructions for their use.

Depending on the testing procedure you use, the dissolved oxygen may be measured in parts per million (ppm), or milligrams per liter (mg/L), or milliliters per liter (mL/L). You should be able to make conversions between each of these with the following information:

$$\textbf{ppm } O_2 = \textbf{mg } O_2 / \textbf{L}$$

$$\textbf{mg } O_2 / \textbf{L x 0.698} = \textbf{mL } O_2 / \textbf{L}$$

From this you can also calculate the amount of carbon fixed in photosynthesis as follows:

$$\textbf{mL } O_2 / \textbf{L x 0.536} = \textbf{mg carbon fixed / L}$$

Procedure

1. Fill 3 of the water sampling bottles with water of the 3 different temperatures provided.

2. Determine the DO of each sample using the technique given to you. Record these values in Table 12.1.

3. On the nomogram of oxygen saturation on page 139, use a straightedge or ruler to estimate the percent saturation of DO in your samples and record this value in Table 12.1. Line up the edge of a ruler with the temperature of the water on the top scale and the DO on the bottom scale, then read the percent saturation from the middle scale.

4. Record your values on the class blackboard and then enter class means in Table 12.1.

Table 12.1: Temperature/DO Data

Temperature	Lab Group DO	Class Mean DO	Lab Group % DO Saturation (from nomogram)	Class Mean % DO Saturation (from nomogram)

5. Graph both the lab group data and class means percent saturation as a function of temperature. For this graph you will need to determine the following:

   a. The *independent* variable: ______________________________
   Use this to label the horizontal (x) axis.

   b. The *dependent* variable: ______________________________
   Use this to label the vertical (y) axis.

Graph 12.1 Title: ______________________________

Figure 12.2: Nomogram of Oxygen Saturation

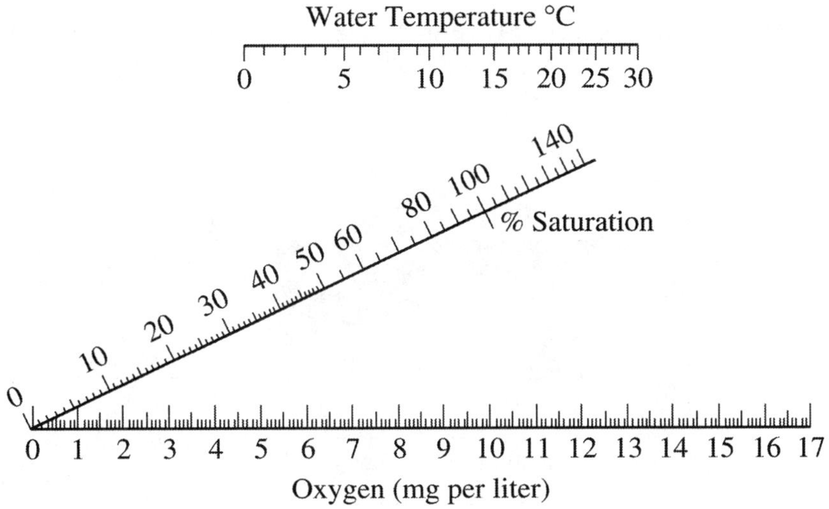

Productivity

The **primary productivity** of an ecosystem is defined as the **rate** at which organic materials (carbon-containing compounds) are stored. Only those organisms possessing photosynthetic pigments can utilize sunlight to create new organic compounds from simple inorganic substances. Green plants obtain carbon for carbohydrate synthesis from the carbon dioxide in the water or the air according to the basic equation for photosynthesis:

$$6CO_2 + 6H_2O \rightarrow C_6H_{12}O_6 + 6O_2$$

The rate of carbon dioxide utilization, the rate of formation of organic compounds, or the rate of oxygen production can be used as a basis for measuring primary productivity. A measure of oxygen production over time provides a means of calculating the amount of carbon that has been bound in organic compounds over a period of time. For each milliliter of oxygen produced, approximately 0.536 milligrams of carbon has been assimilated.

One method of measuring the rate of oxygen production is the **light and dark bottle method.** In this method, the DO concentrations of samples of ocean, lake, or river water, or samples of laboratory algal cultures, are measured and compared before and after incubation in light and darkness. The difference between the measurements of DO in the initial and dark bottles is an indication of the amount of oxygen that is being consumed in respiration by the organisms in the bottle. In the bottles exposed to light, the biological processes of photosynthesis and respiration are occurring; therefore, the change over time in DO concentration from the initial concentrations is a measure of **net productivity.**

The difference over time between the DO concentrations in the light bottle and the dark bottle is the total oxygen production and therefore an estimate of gross productivity (see Figure 12.3).

Figure 12.3: Light-Dark Bottle Method to Determine Gross Productivity

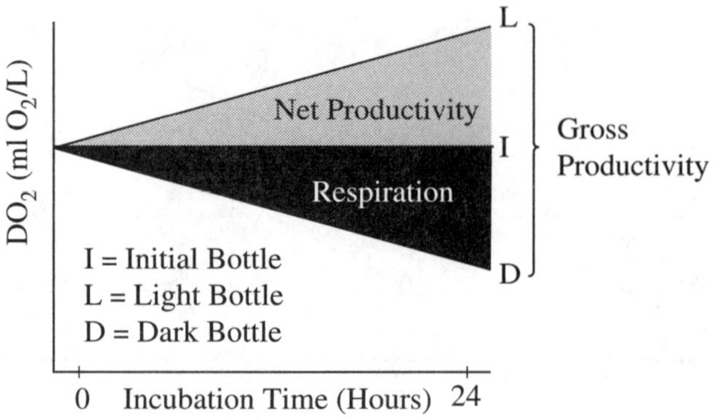

L – I = Net Productivity
I – D = Respiration
L – D= Gross Productivity

EXERCISE 12B: A Model of Productivity as a Function of Depth in a Lake

Day One

1. Obtain 7 water sampling bottles (these are also called BOD bottles, for "biological oxygen demand"). Fill all the bottles with the lake water or algal sample provided. (You may be asked to add a specific weight of aquatic plants to each bottle.) Be careful not to leave any air bubbles at the tops of the bottles.

2. Use masking tape to label the cap of each bottle. Mark the labels as follows: I (for "initial"), D (for "dark"), 100%, 65%, 25%, 10%, and 2%.

3. Determine the DO for the "Initial" bottle now. Record this DO value in Table 12.2 (page 141) and in the data table on the blackboard. Record the class "Initial" bottle mean in Table 12.2. This is the amount of DO that the water has to start with (a base line).

4. Cover the "Dark" bottle with aluminum foil so that no light can enter. In this bottle no photosynthesis can occur, so the only thing that will change DO will be the process of respiration by *all* of the organisms present.

5. The attenuation of natural light that occurs due to depth in a body of water will be simulated by using plastic window screens. Wrap screen layers around the bottles in the following pattern: 100% light — no screens; 65% light — 1 screen layer; 25% light — 3 screen layers; 10% light — 5 screen layers; and 2% light — 8 screen layers. The bottles will lie on their sides under the lights, so remember to cover the bottoms of the bottles to prevent light from entering there. Use rubber bands or clothespins to keep the screens in place.

6. Place the bottles on their sides under the bank of lights in the classroom. Be sure to turn the bottles so that their labels are down and do not prevent the light from getting to the contents. Leave overnight under constant illumination.

7. (Optional Exercise.) If time permits, make a wet mount slide of a sample of the lake water used for this experiment and draw some of the organisms you observe. Can you identify them?

Table 12.2: Respiration

	Individual Data	Class Mean
Initial DO		
Dark Bottle DO		
Respiration Rate (Initial—Dark)		

Day Two

8. Determine the DO in all the bottles that have been under the lights. Record the "Dark" bottle DO in Table 12.2. Calculate the respiration rate using the formula in the table. Record the values for the other bottles in Table 12.3. Complete the calculations in Table 12.4 to determine the gross and net productivity in each bottle. The calculations will be based on a time period of 1 day. Enter your respiration rate and gross and net productivities in the data table on the class blackboard. Determine the class means. Enter these means in Table 12.2 and Table 12.4.

Table 12.3: Individual Data—Productivity of Screen-Wrapped Samples

# of Screens	% Light	DO	Gross Productivity [Light Bottle - Dark Bottle]	Net Productivity [Light Bottle - Initial Bottle]
0	100%			
1	65%			
3	25%			
5	10%			
8	2%			

Table 12.4: Class Data—Mean Productivity

# of Screens	% Light	Gross Productivity	Net Productivity
0	100%		
1	65%		
3	25%		
5	10%		
8	2%		

9. Graph both net and gross productivities as a function of light intensity (class means). The two kinds of productivity may be plotted on the same graph.

For this graph you will need to determine the following:

a. The *independent* variable: ______________________________
Use this to label the horizontal (x) axis.

b. The *dependent* variable: ______________________________
Use this to label the vertical (y) axis.

Graph 12.2 **Title:** ______________________________

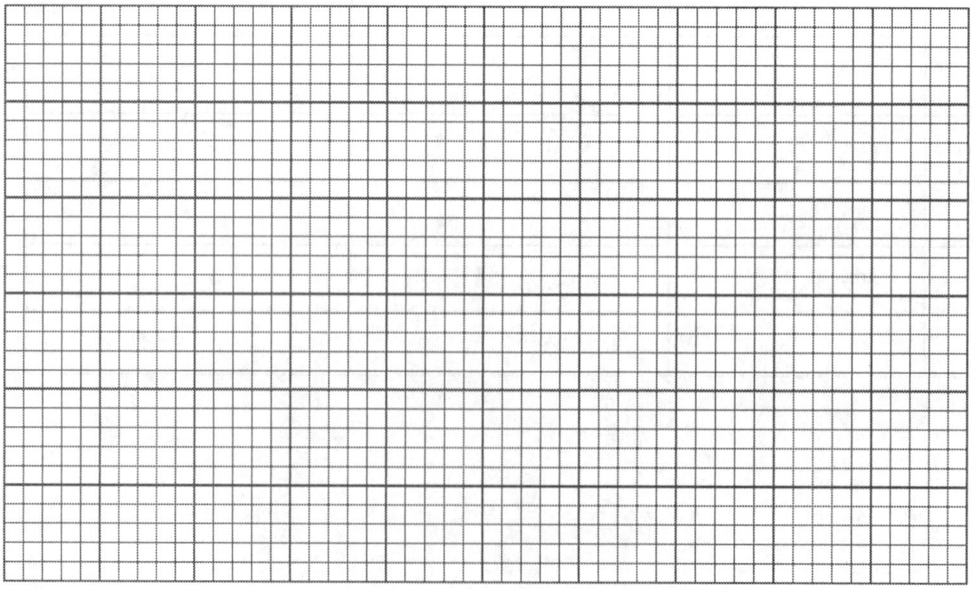

Questions

1. What are three ways primary productivity can be measured?

____

____

____

2. What is the relationship between oxygen production and assimilation of carbon?

____

____

____

3. From your graph of the temperature data, what is the effect of temperature on the amount of oxygen that water at different temperatures can hold?

____

____

____

____

4. Refer to your graph of productivity and light intensity. At what light intensity do you expect there to be:

   No gross productivity?__________ No net productivity? ________

5. A mammal uses only 1 to 2 percent of its energy in ventilation (breathing air in and out) while a fish must spend about 15 percent of its energy to move water over its gills. Explain this huge difference in their efforts to collect oxygen.

____

____

____

____

6. Would you expect the DO in water taken from a stream entering a lake to be higher or lower than the DO taken from the lake itself? Explain.

____

____

____

____

7. Would you expect the DO concentration of water samples taken from a lake at 7:00 a.m. to be higher or lower than samples taken at 5:00 p.m.? Explain.

8. In the following drawings of identical containers with identical fish but with different volumes of water, which one, A or B, would have more oxygen available to the fish? Explain.

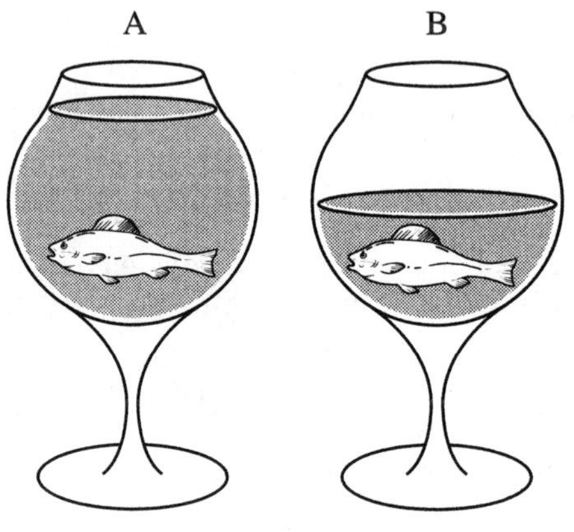

9. What is eutrophication? Research and explain why allowing nitrogen or phosphorous fertilizers to run into a body of water can negatively affect life in it.

APPENDIX I LAB INVESTIGATIONS AND EXERCISES

Labs are an important part of your course work because they are a way to help you apply the themes and concepts you are studying in class to biological processes. Some important goals of lab exercises include helping you to: make systematic observations of the natural world; learn how to use scientific apparatus; make predictions and hypotheses; gather, analyze and interpret data; communicate experimental results; and pose questions for further investigation. They require you to identify assumptions, use critical and logical thinking, and consider alternative explanations.

As an example, suppose you are a member of a group that is investigating the effects of different amounts of nitrogen on the growth of a particular species of plant. How would your investigation proceed to meet the goals given above?

Identify Questions and Concepts

Your group would first define and agree on the problem to be investigated. Questions such as "What percentage of nitrogen would cause the plants to grow the tallest?" help to focus the investigation on one particular aspect of the plant's response to the added nitrogen. Next, you would develop several hypotheses, or educated guesses, that describe the plant's responses. Hypotheses are presented in an if... then... format; for example, "*If* plants are given additional nitrogen *then* they will grow taller than plants that receive no additional nitrogen."

In developing a good hypothesis it might be necessary to conduct a review of textbooks or other resources. For example, a review would help clarify what role nitrogen plays in plant growth.

Design and Conduct the Investigation

Using the hypothesis, the group would next design an experiment to compare the effects of different amounts of nitrogen added to the plants. It would need to identify constants (controlled variables), those factors that remain consistent throughout the experiment. Constants would be such factors as the species of plant used, the volume of nitrogen applied to the plants, the amount of light each plant receives, room temperature, and the form in which the nitrogen is delivered. Also, the group would need to identify the variables. An independent variable is a factor that is purposefully changed, in this case the percentages of nitrogen. There might be several levels of the independent variable, for example, nitrogen concentrations of 1%, 3%, 5%, 7%, and 10%. A dependent variable is the variable that responds to the change in the independent variable. In this experiment, the height of the plant would be the dependent variable.

The experiment would also need to have a control, a group of plants that receive no additional nitrogen (in other words, a group in which the independent variable is not changed). Controls are a way to detect or measure the influence of unanticipated factors, and they establish a base line for comparing the experimental effects to no treatment.

Finally, the group would need to include repeated trials (replication) in the experimental design. These repeated trials reduce the probability that a change occurred because of chance, some random factor, or because of individual variations within the plants.

Use Appropriate Technology

An important component of any scientific investigation is the use of appropriate technology that aids in the collection, analysis, and presentation of data. The group could use computers and spreadsheets to organize the raw data and summarize it by using statistical analysis. Graphs and charts could be produced that would clearly and concisely communicate the results.

Formulate and Revise Explanations and Communicate the Results

The final task for the group would be to analyze the results and communicate them in an accurate and effective manner. This would be accomplished by constructing a reasoned argument, careful and logical writing, and the use of charts and graphs. The discussion might include other plausible explanations, but these would need to clearly identify and support the best explanation of the observed results. Other components of the report might include an explanation of the statistical analysis, an oral presentation, and the appropriate response to critical comments. Discussion questions are an important way of stimulating and directing dialogue about a topic or concept.

Investigate Further

Lab exercises often lead to other questions or suggest alternative methods for performing the investigation. For example, using the goals and steps given above, the group might wish to investigate whether nitrates, nitrites, or ammonia are most efficiently used by the plant; or they might choose to determine if delivery to the roots or the leaves is the best way to get the nitrogen into the plant.

APPENDIX II CONSTRUCTING LINE GRAPHS*

Suppose we are studying some chemical reaction in which a substance, *A*, is being used up. We begin with a large quantity (100 mg) of *A*, and we measure in some way how much *A* is left after different times. The results of such an experiment might be presented pictorially like this:

Figure A.1

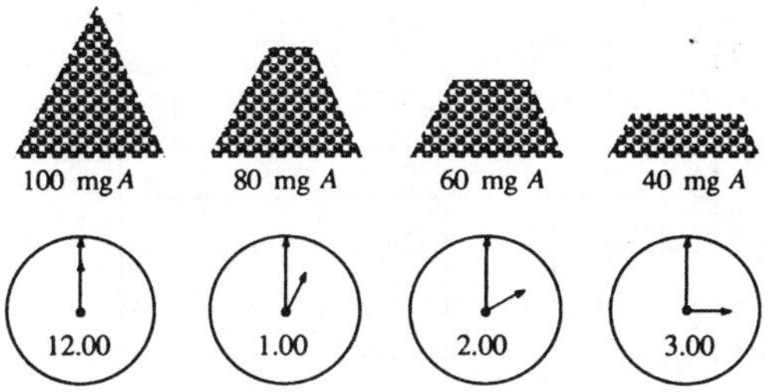

This is the kind of picture graph that you often see in newspapers. This information can be presented much more simply on a graph — a line graph is permissible — because our experience tells us that when *A* is disappearing in a chemical reaction, it is disappearing more or less smoothly and will not suddenly reappear again. In other words, the progress of a chemical reaction is a continuous process, and because time is a continuous process it is permissible to relate the two kinds of information to one another on a line graph. The procedure for constructing the line graph is shown in Figure A.2.

Figure A.2

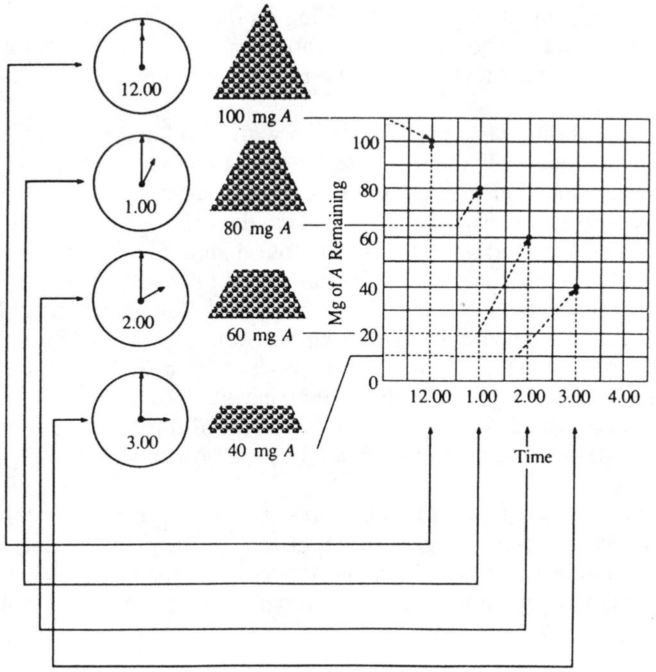

* Based on a handout by Dr. Mary Stiller, Purdue University.

It should be clear from the diagram that each point corresponds both to a particular measurement of the amount of *A* remaining and to the particular time at which that amount remained. (A heavy dot is made opposite both of these two related quantities.) When all the measurements have been recorded in this way, we connect the dots with a line, shown in Figure A.3. (Figures A.21–A.23 explain when to connect the data points.)

Figure A.3

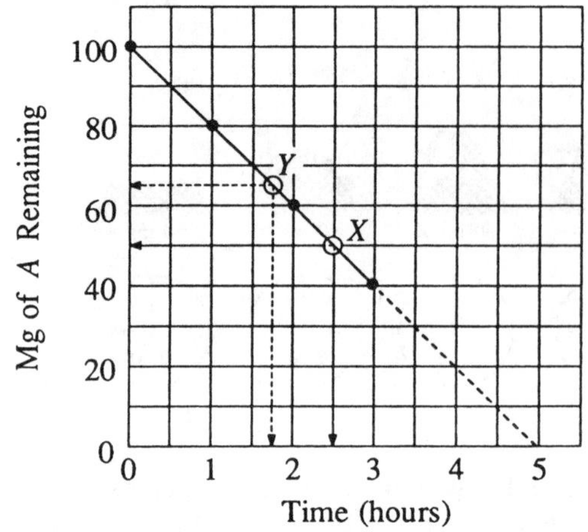

It should be clear by looking at our graph that the only measurements we actually made are those indicated by the dots. However, because the information on both scales of the graph is assumed to be continuous, we can use the graph to find out how much *A* would have been found if we had made our measurements at some other time, say 2.5 hours. We merely locate the line that corresponds to 2.5 hours on our time scale and follow it up until it crosses our line graph at the point X; then we look opposite X to the "Mg of *A* Remaining" scale, and read off 50 mg. We conclude, then, that if we had made a measurement at 2.5 hours, we would have found 50 mg of *A* left. In a similar way, we can find out from our graph at what time a given amount of *A*, say 65 mg, would be left. We have merely to find the line that represents 65 mg on the vertical scale and follow it across until it cuts the line graph at point Y. Then we see 1.75 hours on the "Time" scale opposite Y. This tells us that had we wished to stop the reaction with 65 mg of *A* remaining, we would have had to do so after 1.75 hours.

You will notice that part of the graph has been drawn with a broken line. In making a line graph we are properly allowed to connect only the points representing our actual measurements. It is possible that measurements made after 3 hours will give points that will fall on the broken-line extension of the graph, but this is not necessarily so. In fact, the reaction may begin to slow up perceptibly, so that much less *A* is used up in the fourth hour than in the third hour. Not having made any measurements during the fourth hour, we cannot tell, and we confess our ignorance quite openly by means of the broken line. The broken line portion of the graph is called an **extrapolation,** because it goes beyond our actual experience with this particular reaction. Between any two of our measured points it seems fairly safe to assume that the reaction is proceeding steadily, and this is called an **interpolation.** Interpolations can only be made between measured points on a graph; beyond the measured points we must extrapolate. We know that the amount of *A* remaining after 4 hours is somewhere between 0 and 40 mg. The amount indicated by the broken line on the graph, 20 mg, is only a logical guess.

Unfortunately, it sometimes happens that even professionals take this sort of limitation of line graphs for granted and do not confess, by means of a broken line, the places where they are just guessing. Therefore, it is up to readers of the graph to notice where the last actual measurement was made and use their own judgment about the extrapolated part. Perhaps the extrapolated part fits quite well with the reader's own experience of this or a similar reaction, and he or she is quite willing to go along with the author's extrapolation. On the other hand, the reader may be interested only in the early part of the graph and be indifferent to what the author does with the rest of it. It may also be that the reader knows that the graph begins to flatten out after 3 hours and so disagrees with the author. The point is that we, the readers, must be aware of what part of the graph is extrapolated, that is, predicted, from the shape of the graph up to the time when the measurements were stopped. Hence, you must clearly indicate on a line graph the points that you actually measured. Regardless of what predictions or conclusions *you* want to make about the graph, you *must* give the reader the liberty of disagreeing with you. Therefore, it is very improper to construct a line graph consisting of an unbroken line without indicating the experimentally determined points.

Basic Requirements for a Good Graph

The following procedure applies primarily to graphs of experimental data that are going to be presented for critical evaluation. It does not apply to the kind of rough sketch that we often use for purposes of illustration.

Every graph presented for serious consideration should have a good **title** that tells what the graph is about. Notice that we need more than just a title; we need a *good* title. Before we try to make a good title, let us look at an example and try to decide what kind of title is a useful one. Look at Figure A.4.

Figure A.4

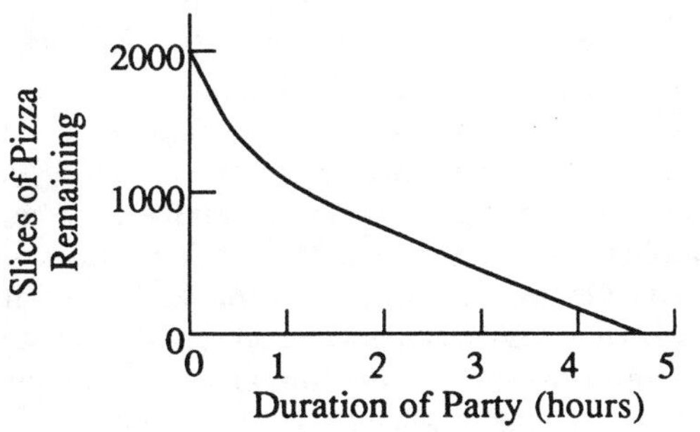

If you like pizza, it might be very useful to know when this party is being held. Without a title, you cannot tell even whether the graph refers to any particular party at all. It might represent average figures for all the parties held last year, or it might represent the expected figures for a party that is going to be held tonight. Let us suppose that these data refer to a study party given by AP Biology students on March 9. Here, then, are some possible titles:

(a) The APs Have a Party

(b) Pizza Rules! Enjoy it with AP

(c) An AP Biofeast!

None of those titles are especially useful or informative, because none of them tell what the graph is all about. Now look at these two titles:

(d) Anticipated Consumption of Slices of Pizza at the AP Biology Party, March 9

(e) Anticipated Consumption of Slices of Pizza at the AP Biology Party, March 9, 1998, 7:00 p.m.–11:00 p.m.

You should be able to see that only title (e) is helpful and useful. It enables you to tell, by glancing at a calendar, whether or not you can attend the party, and it helps make that graph fall a little more steeply. The point we are driving at is that a *good* title is one that tells exactly what information the author is trying to present with the graph. Although brevity is desirable, it should not substitute for completeness and clarity.

Now that you are clear on titles, look at the graph in Figure A.5. Its title tells you that here is some potentially useful information. The graph suggests that, at least for 1998, there was an upper limit to the amount of time people could usefully spend in studying for an exam, and you might wonder, for example, how long you would have had to study to make a perfect score.

Figure A.5: Relation Between Study Time and Score on a Biology Exam in 1998

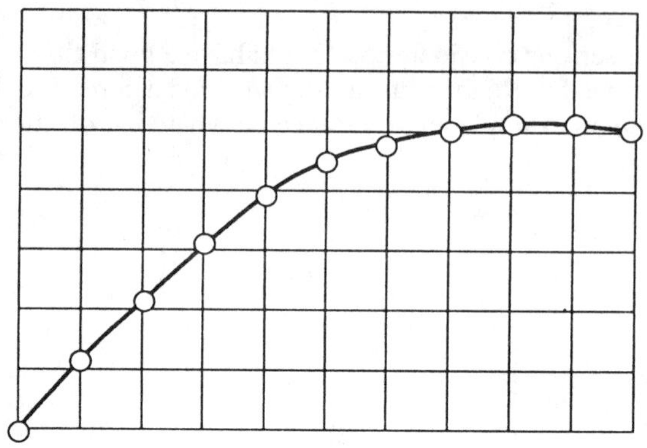

Unfortunately, however, you cannot tell, because the graph has no labels of numbers or units on the scales. Even though this graph has a descriptive and intriguing title, it is of no use to us at all without these very important parts. Obviously, before we can take full advantage of the information that the graph is trying to present, we need to have some additional details.

In Figure A.6 the additional information has been supplied, information that seems to make the graph more useful to us in preparing for the exam.

Figure A.6: Relation Between Study Time and Score on a Biology Exam in 1998

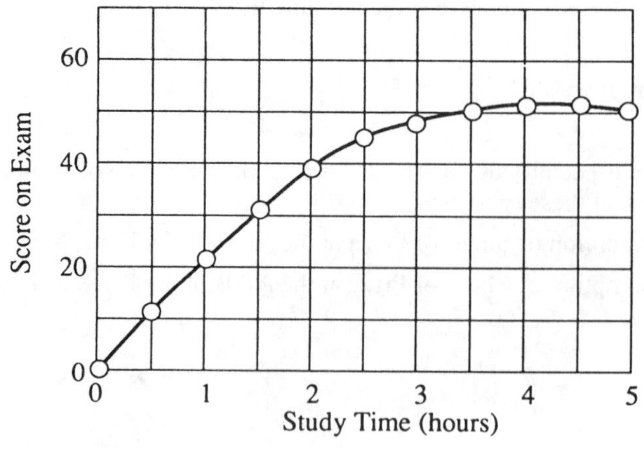

This additional information includes scales, or **axes,** that are carefully marked with numbers, and labels and units that are neatly presented. Obviously, one cannot label all the points along the axes; that would make the numbers crowd together and look sloppy. The units should be marked at intervals that correspond more or less to the intervals between the experimental points. The small marks, called **index marks,** can be drawn in if the experimental points are very widely spaced. Most elegantly, a **frame** is put around the whole graph, and index marks are placed all around. This makes it easy to lay a ruler across the graph when interpolating between the experimental points. The diagram in Figure A.7 summarizes some features of a good graph.

Figure A.7: Relation Between Study Time and Score on a Biology Exam in 1998

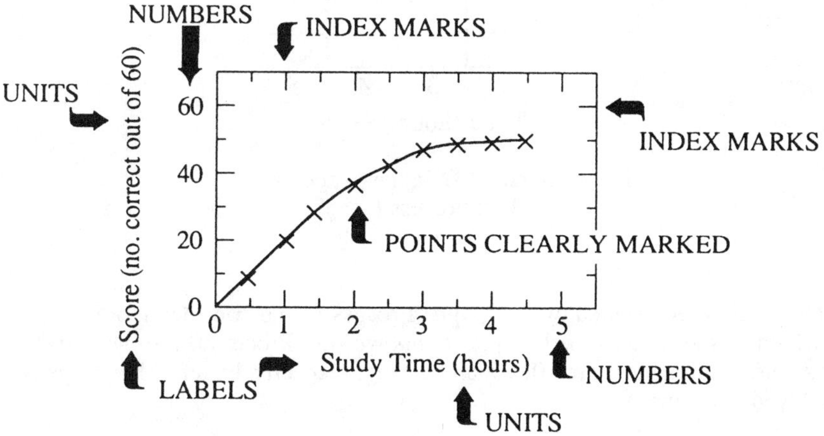

Steepness or Slope of a Line Graph

Look at the graph in Figure A.8 for the disappearance of *A* in a chemical reaction. Such a graph, in which the amount of some quantity is shown on the vertical scale, or **ordinate,** with time shown on the horizontal scale, or **abscissa,** is frequently called a "progress graph" or "progressive curve," because it shows how some process progresses in time. This graph may also be called a "time course" for the process, because it shows the extent to which the process has occurred at different times.

Figure A.8

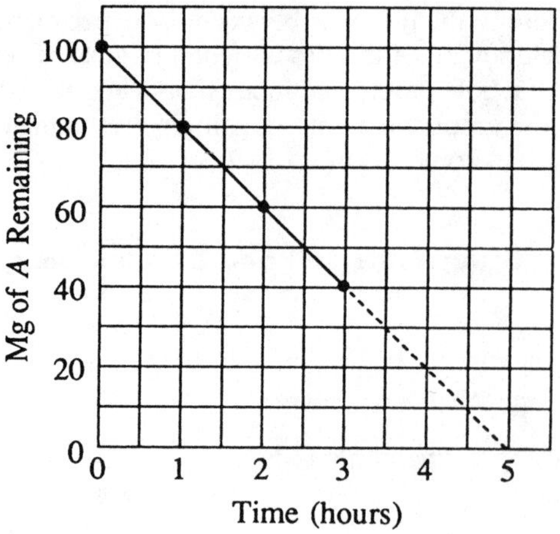

Time Course of Disappearance of
A in Process I

Let us call the process represented by the graph "Process I" and consider another reaction, "Process II," in which *A* is also consumed. Suppose that we start Process II also with 100 mg of *A*, and that after 1, 2, and 3 hours there are 90, 80, and 70 mg, respectively, left. The progress curve for Process II is displayed in Figure A.9.

Figure A.9

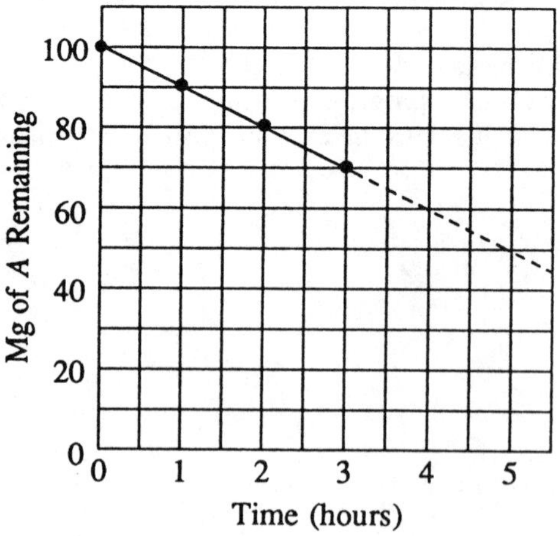

Time Course of Disappearance of
A in Process II

Now, suppose we want to compare the graphs for the two processes. Because they have exactly the same scales, we can put both lines on the same graph, as shown in Figure A.10. Notice, however, that now in addition to the labels on the scales, we need labels on the two lines to distinguish between the two processes.

Look at the 1-hour mark on the time scale of the graph. Opposite this put an X on the line for Process I and a Y on the line for Process II. Then, opposite X on the ordinate you should be able to see that 80 mg of *A* are left in Process I; opposite Y you can see that 90 mg of *A* are left in Process II. Apparently, Process I has used up 20 mg of *A* and Process II has used up only 10 mg in the same amount of time. Obviously, Process I is faster, and the line graph for Process I is steeper than the graph for Process II.

Figure A.10

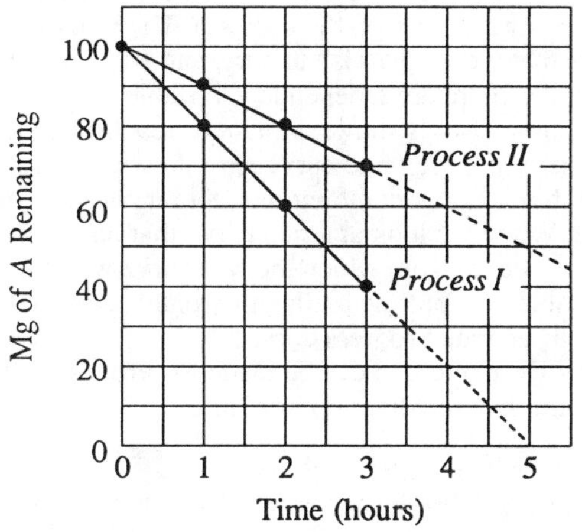

Time Course of Disappearance of *A* in Processes I and II

The rate for Process I is 20 mg *A* used/hr, while the rate for Process II is 10 mg *A* used/hr.

We have seen that a steeper line graph means a faster reaction when the progress curves for two reactions are plotted on the same scale. (Obviously, if the progress curves are plotted on different scales, we cannot compare the steepness of the line directly, but have to calculate what the slope would be if the two curves were plotted on the same scale.)

Suppose, now, that we make a new kind of graph, one that will show the steepness, or slope, of the progress curve. Because the **slope** of the progress curve is a measure of the speed of velocity, or **rate** of the reaction or process, such a graph is frequently called a "rate graph" or "rate curve." The diagram in Figure A.11 shows how a rate curve can be made for Process I.

Figure A.11

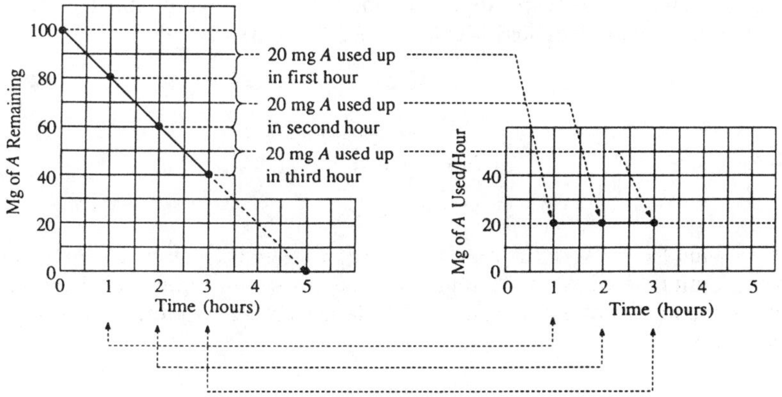

Notice that the time scale of this rate graph is exactly like the time scale of the progress curve from which it was derived, but that the ordinate is different. The ordinate of the progress curve shows milligrams of *A* remaining; the ordinate of the rate curve shows milligrams of *A* used per hour. Obviously, a rate graph must always show rate on one of its scales, and it is ordinarily the vertical one that is used. This is because the rate of a reaction or process is what mathematicians call a **dependent variable.** Time is the **independent variable** in this experiment; it is independent of changes in the dependent variable (the rate of reaction), and it is the variable that is shown on the horizontal axis. Regardless of whether the process is the increase in height or weight of a plant, or the using up or producing of something in a reaction, the rate graph for the process must always show *amount of something per unit time* on one of its axes. One very common type of rate graph is the one shown in Figure A.11, with a rate on the ordinate and the time on the abscissa. Other kinds of rate graphs may have temperature or molarity on the abscissa. The rate of growth of a plant, for example, depends on many factors that we might wish to vary, and so we can have as many different kinds of rate graphs for that process as there are independent variables.

Let us emphasize: a progress curve always shows amount of reaction on the vertical scale and time on the horizontal scale. The corresponding rate curve *may* show time or some other variable on the horizontal scale, but it *always* shows rate, or amount of reaction per unit time, on the vertical scale. This point is very important. When we look at a rate curve that has time on the horizontal scale, we must visualize the progress curve from which the rate curve was derived. When we look at a rate curve that has any other variable except time on the horizontal scale, we shall see that each point on that rate curve represents a separate progress curve.

In the same way as for Process I, a rate curve can be made for Process II. Plotted on the same graph, the two should look something like the diagram in Figure A.12.

Figure A.12

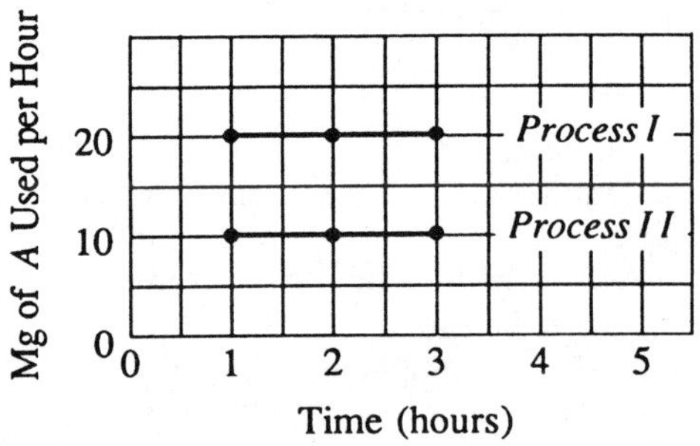

There are two things to notice in this example. First, the curve for Process I lies higher than that for Process II. This is in accord with the facts as we have seen them, namely, that Process I is faster and so has a greater slope or higher value for the steepness. Second, notice that both curves are perfectly flat. Naturally, because the progress curves for the two processes were both perfectly straight lines, having everywhere the *same* slope, the rate of steepness graph must show exactly the same thing, that is, that the rate or steepness is everywhere the same.

On the other hand, consider the graph in Figure A.13, which represents the disappearance of *A* in yet another reaction, Process III.

Figure A.13: Time Course of Disappearance of *A* in Process III

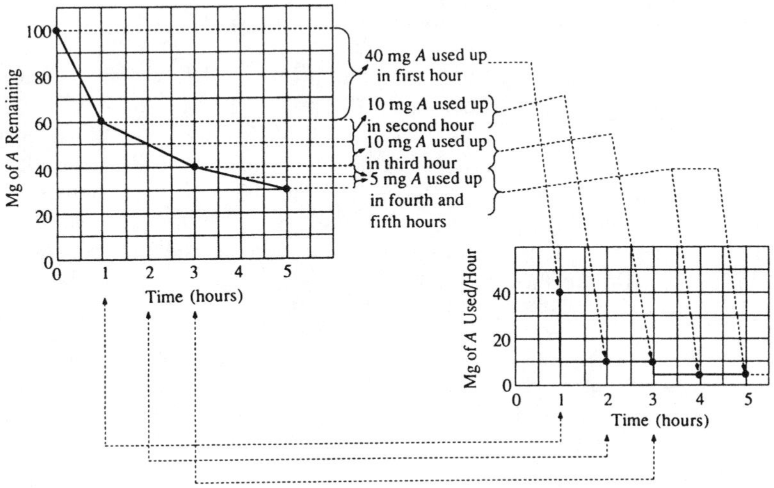

You can see that Process III differs from Processes I and II in that the progress curve for III is not a perfectly straight line. It is steepest at the beginning, becomes less steep after 1 hour, and again after 3 hours. Obviously, because the rate of the process is changing with time, the corresponding rate curve will *not* be perfectly flat. The rate has to start out high, then drop at 1 hour and at 3 hours, and you can see in the graph on the right that this is exactly what it does. In fact, the rate curve looks like steps, because whenever the slope of the progress curve decreases, the rate curve must show a drop to a lower value. Conversely, if the progress curve for a process should get steeper, as sometimes happens (the reaction goes faster after it gets "warmed up"), the rate curve must show a corresponding increase to a higher value.

Until now we have been able to read the steepness, or slope, of the progress curve directly from the scales of the graph, because the progress curves we have been studying were either perfectly straight lines or else made up of straight-line segments. In most real situations, however, we cannot do this, because the slope of the progress curve does not change sharply at a given time, but gradually, over a period of time. You probably remember how to measure the slope of a curved line, but let us review the process anyway. (See Figure A.14.)

Figure A.14

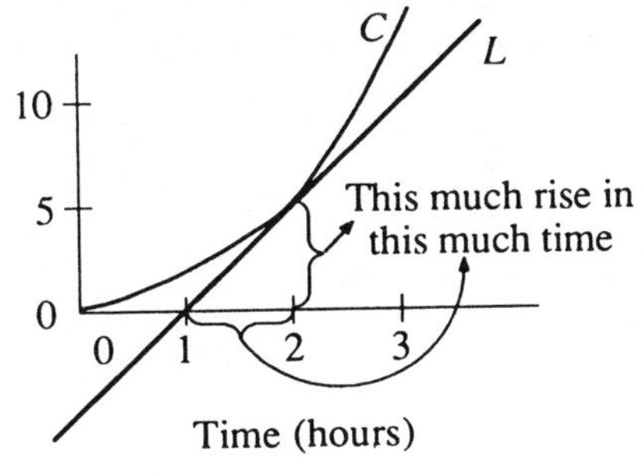

Suppose we want to measure the slope, or steepness, of the curved line C at time 2 hours. We can see that the curve rises 5 units total in the 2 hours, so that the average slope is 2.5 units per hour. However, it is easy to see from the graph that this average is very misleading; the progress curve is almost flat at the beginning (i.e., has 0 slope) and then accelerates rapidly, so that the line curves upward. If we want to find the true slope at 2 hours, we must draw line L in such a way that L has the same slope as C at the 2-hour point. Then we can see that L rises about 5 units between 1 and 2 hours, just twice the average slope for the first 2 hours.

We have seen that a perfectly flat rate curve, like that for Process I or II, means that the corresponding progress curve is a perfectly straight line having the same slope at all points. Conversely, a progress curve that changes in slope, like that of Process III, will give a rate curve that looks like steps. You should be able to figure out that the "steps" on the rate curve will be sharp and square if the progress curve has an abrupt change in slope, and more rounded off if the progress curve changes slope gradually. In any case, in regions where the rate curve is perfectly flat it is clear that the progress curve must have constant steepness, or slope. However, if the progress curve itself gets perfectly flat, then that portion of the progress curve has 0 slope; in other words, the reaction has stopped. This kind of situation is pictured in Figure A.15 where the rate and progress curves for another reaction, call it Process IV, are shown.

Figure A.15

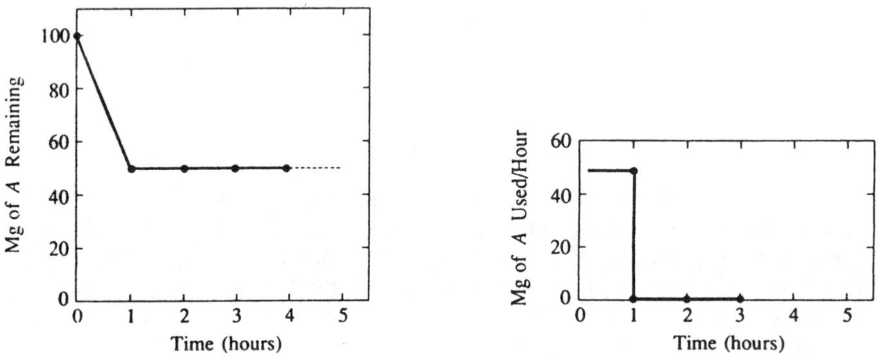

Progress Curve

Rate Curve for the Disappearance of A in Process IV

In the progress curve on the left, we can see that after the first hour the reaction stopped. From the graph we can see that after 1 hour there were 50 mg of *A* remaining; after 2 hours there were still 50 mg remaining; and there are still 50 mg remaining even at 4 hours. Obviously, Process IV stopped when one-half of *A* had been used up. Now look at the rate curve on the right. It is perfectly flat for the first hour, because the slope of the progress curve was constant during that time. After the first hour the rate curve is also perfectly flat but it has dropped down to 0, indicating that although the progress curve has constant slope, the slope is actually 0. Obviously, flatness in a rate curve and flatness in a progress curve mean different things. Flatness in the progress curve for a reaction means that the reaction has stopped; flatness in the rate curve means that the reaction is going on at a constant rate. You can see, then, that we have to be able to glance at a graph and tell whether it is a rate curve or a progress curve in order to be able to interpret what the shape of the curve is trying to tell us.

Now let us take one more example of this kind of rate curve. The graph in Figure A.16 shows the progress in the growth of a pea plant. First, we can see that the slope is not the same everywhere. In fact, there is an interval where the slope increases very gradually from 0. By 1 week or so the slope has reached its maximum value and is steady until about 3 weeks. Thereafter, the slope begins to decrease again, as the curve bends over, and eventually, at about 4.5 weeks, as the curve gets perfectly flat, the slope, or steepness, tends to be 0 again.

Figure A.16

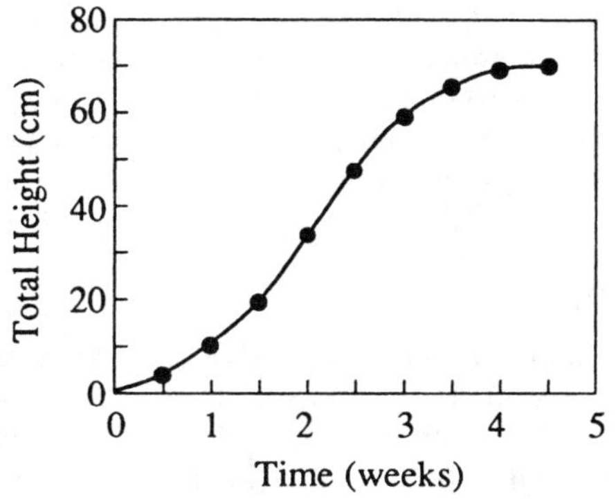

Progress Curve for the Growth of a Pea Plant

Suppose, now, that we try to imagine what the rate curve for the growth of this pea plant will look like. If you read through the preceding paragraph, you will have a rough description of it. In fact, it will look like the graph in Figure A.17.

Figure A.17

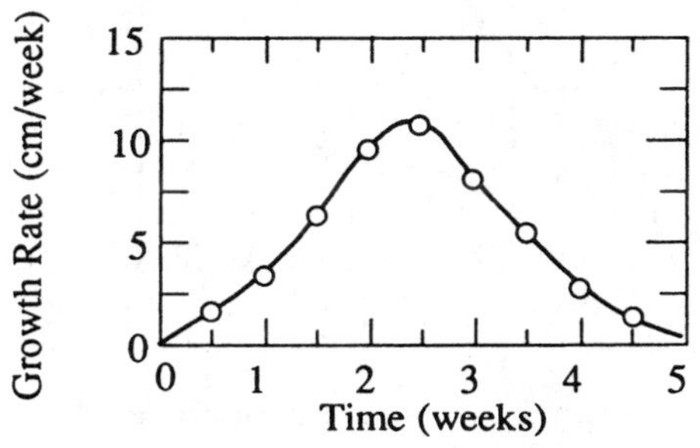

Growth Rate of a Pea Plant

Notice from the two graphs that where the steepness of the progress curve gets larger, the corresponding rate curve turns upward. Similarly, when the slope of the progress curve decreases again, the rate curve turns downward. A rate curve that is turning up means, therefore, that the process is speeding up; a flat rate curve means that the process is going at a constant rate; and a rate curve that is turning down means that the process is slowing down. When the rate curve hits the x-axis, it means that the reaction has stopped.

Probably 80 percent of the graphs you will encounter in biology are either rate curves or progress curves. You will have noticed from the preceding discussion that biologists tend to use the words "graph" and "curve" interchangeably. Technically, of course, the entire picture, including the abscissa, ordinate, labels, numbers, units, index marks, and title, together with the line graph portrayed, is the "graph," while the line graph itself is called the "curve." You will notice, too, that biologists call a line graph a "curve," even though the line itself may be perfectly straight.

To summarize, remember that a progress curve is made from measurements at different times during the progress of a reaction that is continuous with time. A graph that shows how much or to what extent a reaction has occurred at different times is a progress or time-course curve. In contrast, a rate curve is a picture of the steepness of one or more progress curves, and any graph that has rate on one of its scales is a rate curve.

So far we have been considering only rate graphs that have time on the abscissa; we could call these **time-rate** curves. As we have seen, a time-rate curve can be made from any progress curve. Next, we are going to consider rate curves that do not have time as the abscissa. As you shall see, such curves are made by combining data from several progress curves, each representing the time course of the reaction under a different set of conditions.

Other Kinds of Rate Graphs

Let us look at and try to analyze the graph in Figure A.18. Obviously, it is a progress curve, because it shows an amount of something on the ordinate and time on the abscissa. There are several different curves all plotted on the same graph, and each is labeled with a different temperature. The title indicates that this graph is trying to tell us how Process I behaves at different temperatures.

Figure A.18

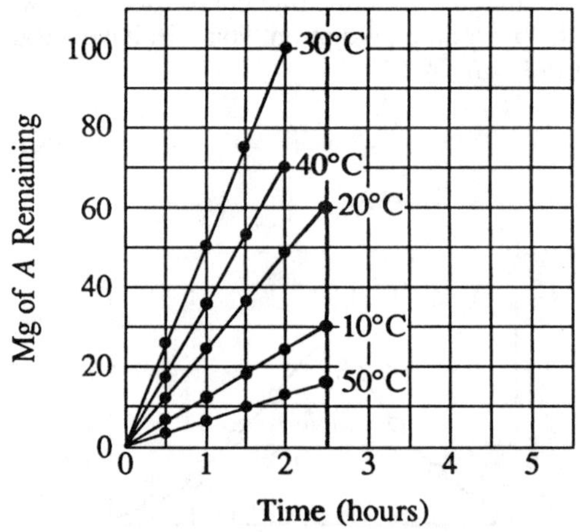

Temperature Dependence of Process I

Before we try to construct the rate curve for this graph, we should try to imagine how this experiment was carried out. It seems clear that the experiment must have started with several different batches of *A* and that each reaction mixture was kept at a different temperature. Then, every half-hour, the amount of *A* remaining was measured and the amount consumed was calculated. The results might have been plotted in five separate progress curves, as shown in Figure A.19.

Figure A.19

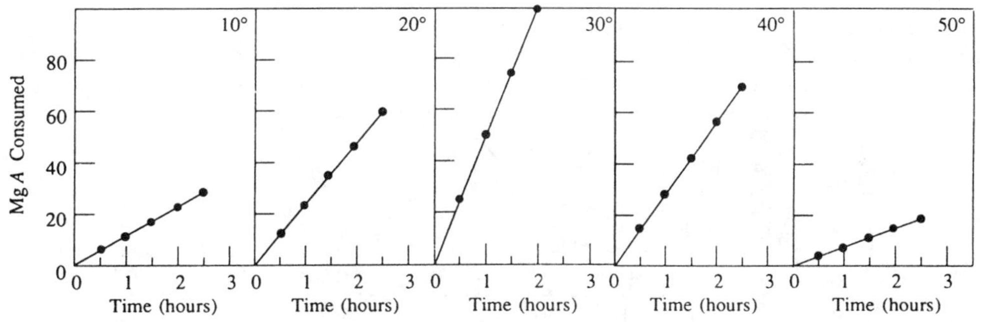

When all these progress curves are plotted on the same graph, as was done in Figure A.18, we have what is called a "family" of curves. If we look at the slopes of the various members of the family of curves for Process I, we see that the steepest slope does not correspond to the highest temperature. In fact, the curve for 30° is the steepest, whereas the curve for 50° is the least steep; the curve for 10°, the lowest temperature, has an intermediate slope. By analyzing and comparing the slopes of the family of curves in this way we can get a reasonably good notion of the effect of temperature on Process I, but this effect could be shown much more clearly in a rate graph that has temperature as the abscissa. Such a graph would show us at a glance how the rate varies with temperature and, of course, would be preferable, as the whole point in making a graph is to present information simply and clearly. The diagram in Figure A.20 shows how a **rate-temperature** graph would be constructed from this family of curves for Process I.

Figure A.20

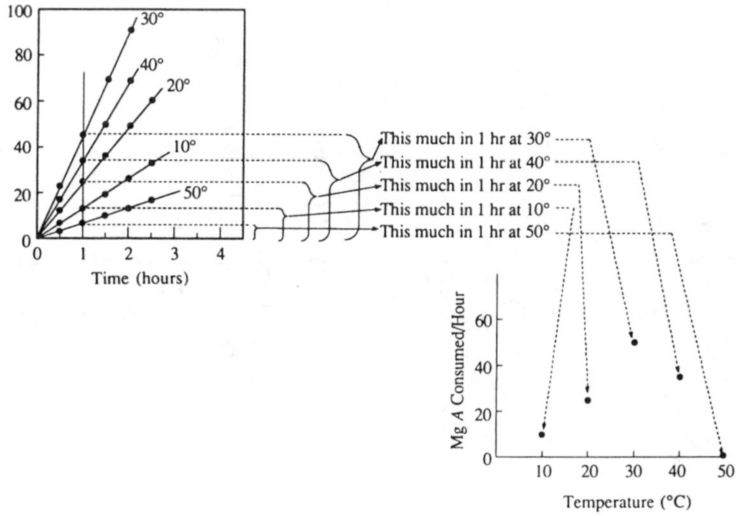

Having found, as shown in Figure A.20, the five points for our rate graph, we are faced with the question of whether or not it is legitimate to connect these points with a smooth line. We recognize, of course, that both temperature and rate are continuous processes. Between any two given temperatures or rates there are an infinite number of temperatures or rates. The question here, however, is the following: If we do draw a smooth line through our five points, will that line pass through the infinite number of other rates that we could have measured if we had chosen some other temperature? Let us go ahead and draw the line, as shown in Figure A.21.

Figure A.21

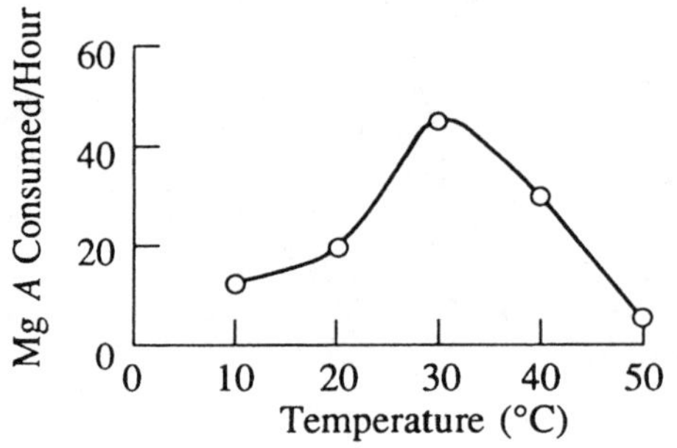

As we have drawn it, the curve indicates that the rate at 29° and at 31° would be slightly lower than at 30°, and this may not be true. In order to determine the true shape of the curve in the region of the maximum rate we would have to make progress curves at smaller temperature intervals, say, every two degrees. However, it is extremely unlikely that the true shape of the curve is anything like the two possibilities shown on the diagrams in Figure A.22. All our experience tells us that if a reaction depends on temperature, then that dependence will be a smooth curve, without sharp bends. In fact, if in an experiment we *should* observe behavior of the type shown in Figure A.22, we would immediately begin to suspect that something is wrong with our thermostat! Thus, although it may be that the shape of the rate-temperature curve for Process I is somewhat different from the way we drew it in Figure A.21, we can be reasonably certain that it is not radically different.

Figure A.22

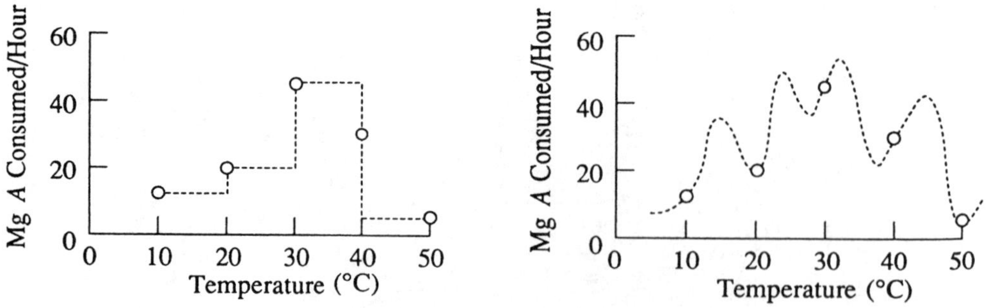

In addition, we may also tend to be suspicious of a graph if we see a sharp peak, unless the experimental points were taken very close together. For example, common sense would tell us to be careful about accepting the rate curve shown in Figure A.23.

Figure A.23

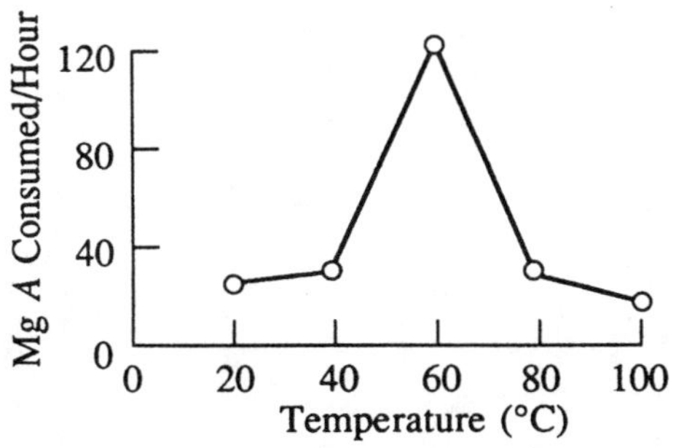

Obviously, most of the shape is given to the profile by the one measurement at 60°. In biology, as in everything else, mistakes can be made, so the experimenter would have to check the validity of that measurement very carefully. The easiest way to do that would be to make more measurements slightly above and slightly below 60° to see whether these would fall on the line the experimenter has drawn. Alternatively, the experimenter could play it safe and draw only a bar graph for these spaced temperatures. Another useful dodge would be to connect the points with a smooth but broken line rather than a continuous line. As always, the broken line would suggest the tentative and provisional nature of the curve as drawn.